God Made
PLANTS

Science/Worldview | Level 2

Author: Tamela Sechrist

Editors: R.A. Sheats, Kayla White

Generations
PASSING ON THE FAITH

God Made
PLANTS

Hi! We would like you to be your outdoor friends! We're very excited about plants. God made many kinds of plants and gave them special jobs. I'm Herby and I like to work!

And I'm Flora! I'm so happy to be a beautiful flower that shows God's beauty. Let's learn about plants together!

Printed in the United States of America.

ISBN: 978-1-7350719-9-2

Cover Design: Justin Turley
Interior Design: Sarah Lee Bryant

Published by:
Generations
19039 Plaza Drive Ste 210
Parker, Colorado 80134
Generations.org

For more information on this and other titles from Generations,
visit Generations.org or call (888) 389-9080.

Table of Contents

Introduction

This introductory science course for young, school-age children is designed to bring to light the love, wisdom, and power of God that is evident in His creation. *God Made Plants* presents the amazing way that plants are perfectly designed to provide the earth with food, oxygen, moisture, healthy soil, and materials for supplies. There would be no life on Earth without plants!

Herby, the hardworking leaf, appears often in the course, both to keep the child's interest and to rejoice in the hard work plants do. Flora, the lovely flower, also has praise for God's plants—especially their beauty!

God Made Plants is designed to work with the way God made children to learn. Children learn best by using a variety of their senses and through hands-on activities. This curriculum helps children learn through:

- **Listening:** The textbook is designed to be read aloud to the child. The language is simple and easy to understand. Important words are in **bold** so the reader knows to emphasize them and maybe take a detour to a definition box or picture.

- **Seeing:** The beautiful pictures, fun illustrations, and simple diagrams in the textbook provide a visual aspect to learning.

- **Moving:** Children will have an opportunity to grow their own garden and watch God's amazing plan for plants firsthand! In addition, the *God Made Plants* companion activity book is full of hands-on learning, involving the senses and muscles with a variety of fun activities that are not too burdensome for the parent/teacher. The science topics are reinforced with a balance of observation, experiments, imagination, logic, Scripture, art, cooking, poetry, math, exercise, music, geography, and a little bit of writing. Each activity reinforces the material introduced in the textbook, and every exercise is numbered for easy reference.

How To Use *God Made Plants*

God Made Plants is divided into nine units of four chapters each. Units 1-8 each have three chapters of instruction on an aspect of plants, followed by a fourth chapter on a biome of the world. For example, chapters 1-3 cover the wonderful things plants do and provide for us. Then, chapter 4 covers the grassland biome: its plants, animals, and location in the world. Units 1-8 each have a memory verse and a children's hymn that children will have an opportunity to work on in the activity book.

Hi Parents! Be sure to check out Unit 9, "You Can Plant a Garden," on page 299 to schedule your child's gardening activities for the year!

Unit 9 of *God Made Plants* will guide you and your child in making a small garden. It's recommended that the parent becomes familiar with Unit 9 before starting the course. A garden takes many months to grow from bare soil to harvest. Now and then, you may need to interrupt your progress through Units 1-8 of this book in order to work on Unit 9's garden. Or you may want to keep a steady schedule through the book and work on your garden after school and on weekends. See the introduction to Unit 9 for a schedule of when you may need to allow gardening time throughout the year.

The *God Made Plants* books are organized in a way to enable children to internalize what they have learned. Learning comes easiest in small doses with time between each session. Children solidify what they have learned as they play. It becomes permanent as they sleep. To enable this, the following schedule is suggested for Units 1-8:

Cover one chapter per week for 32 weeks. Each week:

- **Day 1** — Read aloud the first section of the textbook chapter. Complete the corresponding activity in the Activity book (as announced at the end of that section of the text).

- **Day 2** — Break

- **Day 3** — Read aloud the second section of the textbook chapter. Complete the corresponding activity in the Activity Book.

- **Day 4** — Break

- **Day 5** — Complete the Green Thumb Activity from the activity book. This day has no read-aloud section in the textbook. Instead, there is time allowed for a special activity. These Green Thumb activities take a little more preparation and materials than the week's earlier activities.

May God be glorified, and may you be richly blessed as you study His creation in *God Made Plants*!

Tamela Sechrist
The Generations Curriculum Team
June 2022

UNIT 1
God Fills the Earth with Plants

God is very good! In fact, as our memory verse says, the earth is full of the goodness of the Lord. One way God is good is that He fills the earth with life-giving plants. Each of these plants have important jobs to do.

We also see God's goodness to us in the beauty He has made all over the earth. A lot of that beauty is seen in His plants. Green forests, colorful fields, and luscious fruit salads are good things we can praise Him for!

And the earth is full of God's goodness because He sent His Son to the world He loves.

 ## Memory Verse

The earth is full of the goodness of the LORD. (Psalm 33:5)

 ## Hymn to Sing

You can listen to this hymn by searching for "I Sing the Mighty Power of God" on the internet.

I Sing the Mighty Power of God

I sing the mighty pow'r of God, that made the mountains rise,
That spread the flowing seas abroad, and built the lofty skies.
I sing the wisdom that ordained the sun to rule the day;
The moon shines full at His command, and all the stars obey.

I sing the goodness of the Lord, who filled the earth with food,
Who formed the creatures through the Word, and then pronounced them good.
Lord, how Thy wonders are displayed, where'er I turn my eye,
If I survey the ground I tread, or gaze upon the sky.

There's not a plant or flow'r below, but makes Thy glories known,
And clouds arise, and tempests blow, by order from Thy throne;
While all that borrows life from Thee is ever in Thy care;
And everywhere that we can be, Thou, God, art present there.

CHAPTER 1
God Fills the Earth with Food

[God] causes the grass to grow for the cattle,
And vegetation for the service of man,
That he may bring forth food from the earth.
(Psalm 104:14)

How Can the Earth Be Full of Food?

All over the earth, God has created a way for food to be made. How does He do this? He uses three things: **sunlight**, **water,** and **air**. We can find these three things in almost every place on Earth.

Sunlight, water, and air are so

This verse says God brings food out of the earth by making plants (vegetation) to grow!

common that we hardly notice them. But they are also mysterious things. They seem invisible because we can see through them. They also can go almost anywhere, and we have a hard time capturing them. Yet God invented a way that these three mysterious things can be made into good, solid food that we can see, grab, and eat. He created plants to do this job!

The way plants do this is one of the most complicated **processes** on Earth. This process is called **photosynthesis**. We will learn more about it later!

Who gets this food made by plants? God gives this food to everybody and everything that's alive:

1. **People** eat plants. We also eat animals that have eaten plants, and we eat some animal **products**. A product is something that comes from something else or is made out of something else. Milk is a product that comes from cows. Eggs are products that come from chickens. Tortillas and popcorn are products that are made out of corn.

Definitions

A **process** is the step-by-step way that something gets done. A recipe can tell you the process of making a cake. Instructions can tell you the process of putting something together.

Photosynthesis is the process plants use to make food out of sunlight, water, and air.

God fills the earth with food by providing sunlight, water, and air everywhere!

2. **Plants** use the food they make. This food helps them grow. Plants also store food in their seeds to give baby plants energy as they start to grow. Often plants will store food in their roots to use later.

3. **Animals** also eat plants, or they eat other animals that have eaten plants. No people or animals could live if God didn't make plants grow on Earth.

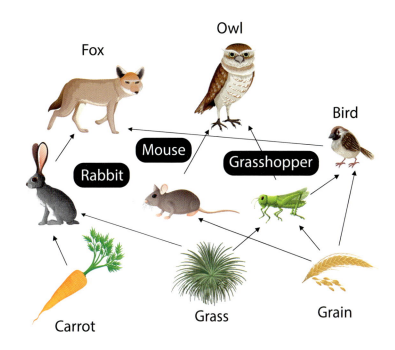

Food Chain

 Time to do Activity 1 in the Activity Book!

What Is This Food?

There are fifty thousand different edible plants in the world! These plants include trees, bushes, vines, and short plants like wide-leafed garden plants and narrow-leafed grasses.

Most of the world's food for people and animals comes from grasses. Grasses make up about one fourth of the plant life on Earth. Different kinds of grasses grow all over the world.

Many kinds of grass make edible seeds, called **grain**, to feed people. Wheat, rice, and corn are grains we

Antarctic hair grass grows near the cold South Pole.

often eat. Grain is a good food because it gives quick energy.

Grain is also good because each plant

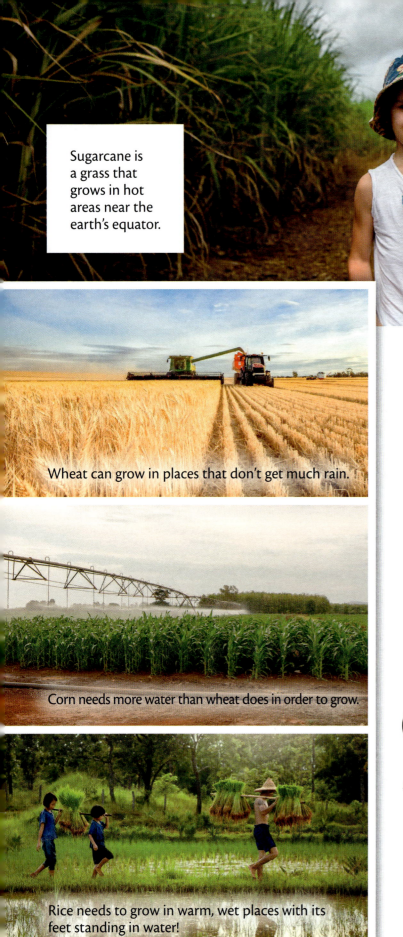

Sugarcane is a grass that grows in hot areas near the earth's equator.

Wheat can grow in places that don't get much rain.

Corn needs more water than wheat does in order to grow.

Rice needs to grow in warm, wet places with its feet standing in water!

can make a lot of food. One grain (or seed) of wheat can grow into a plant that makes about 300 more grains of wheat for us to eat. One grain of rice can be planted and makes about 1000 more grains. One corn seed (kernel) can make about 2000 more! More than half of the world's food for people comes from grain.

Different places on Earth have different amounts of heat, cold, and water, but God has created different grains that are perfect for these different

Definitions

Pastures are fields covered with grass and other plants that are good for animals to eat.

Nutritious food has everything needed for the life and health of the person or animal eating it.

places. Most areas on Earth are blessed with a perfect fruitful grain for the people living there.

People have made good use of grass as food for their animals. Besides letting

After tractors cut hay, it's dried and shaped into round or rectangular bales.

Corn is used to make tortillas in many parts of the world.

their animals eat the grass in **pastures**, people often mow extra grass to dry and store as **nutritious** hay.

In parts of the world where grains do not grow well, God has given people different quick-energy foods from the underground parts of some wide-leafed plants. Potatoes, yams, sweet potatoes, cassava, and taro are plentiful foods for many people in Africa, South America, and on tropical islands.

Quick energy comes from the **carbohydrates** in food. Carbohydrates are sugars and starches. But people need more than quick energy. We also need energy that lasts longer. God has made

Grass is a wonderful creation from God! He made it very fruitful. Grass gives food to people and animals all over the earth.

this possible by giving us **protein** and **fats** in our food.

Many plants we eat have protein and fats. But God has also given us protein and fat from animals. People and many animals can't digest grass. But God gave **grazing** animals (animals that eat grass in a field) a special way to digest grass and use it for food in their bodies. Then we can get protein and fat when we eat

Carbohydrates   **Proteins**   **Fats**

the meat, milk, and eggs from these animals. This is another way God fills the earth with food.

People and animals need more than carbohydrates, protein, and fats from food. Nutritious foods also have important things like vitamins and minerals. God made our bodies to need all these things in order to live. He also made plants that contain these things we need. It's wonderful that He put the right nutrition in plants all over the world for people and animals that live all over the world! He fills the earth with food!

Children help dig taro roots.

This cow's body is turning grass into an udderful of milk.

Prayer

Thank You, God, for growing plants everywhere on Earth! Thank You for putting sunlight, water, and air everywhere. It reminds us that You are everywhere even though we don't see You. You have worked everything perfectly for Your creatures by giving us food this way. Amen.

Time to do Activity 2 in the Activity Book!

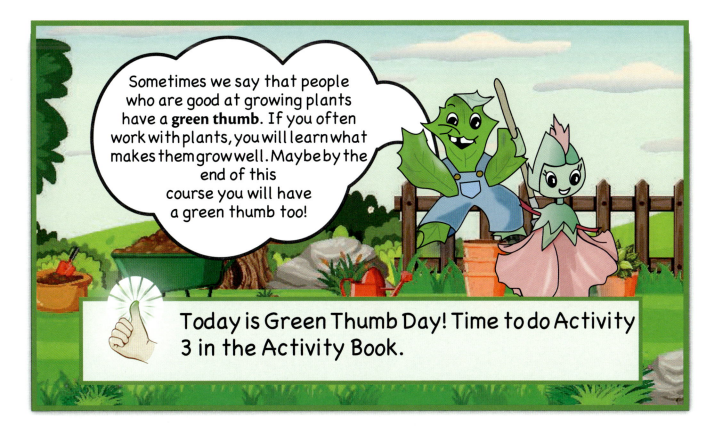

Sometimes we say that people who are good at growing plants have a **green thumb**. If you often work with plants, you will learn what makes them grow well. Maybe by the end of this course you will have a green thumb too!

Today is Green Thumb Day! Time to do Activity 3 in the Activity Book.

CHAPTER 2

God Fills the Earth with Supplies

Plants Are Perfect Gifts Because They Have Structure

We know that God gave us plants to eat as the perfect gift of food. But God also gave us plants to be used as supplies to make things with. We can use plants to make many things. What do plants have that make them perfect as supplies? Every plant, anywhere in the world, has **cellulose**.

Cellulose is what gives plants their **structure**. Cellulose is arranged in different ways to make different parts on the same plant: stems, roots, leaves, and flowers. Cellulose can also be arranged differently to make all the different

Every good gift and every perfect gift is from above, and comes down from the Father of lights. (James 1:17)

Definitions

A plant's **structure** is the way it's put together to give it shape and strength.

Cellulose is tiny strings of sugar, attached and woven together. Cellulose gives plants structure.

See what the Bible says? Good things and things that are perfect for us are gifts from God!

Herby's List of Supplies Made of Cellulose

Paper

Rope

Cardboard

Lumber (wood)

Baskets

Cotton cloth

kinds of plants: skinny grasses, wide-leafed herbs, tiny mosses, huge trees, stiff shrubs, and twisty vines.

Let's look at why cellulose is useful to us:

1. Cellulose is as strong as steel when we bend or stretch it. But it's also lighter than steel. The cellulose in the wood of trees is joined together in a special way to make the trees strong, tall, and sturdy. That's why wood is a perfect gift from God for building houses and furniture all over the world.

2. Cellulose lasts a long time. It can't dissolve in water. It won't rot easily. Cellulose only breaks down if certain creatures, too tiny to see, eat it. People make washable clothes, towels, and bedding out of cotton and linen. Cotton and linen are cellulose. Baskets, rope, paper, cardboard, and wooden things can last a long time because they are made of cellulose.

3. When people eat plants, the cellulose is not digested. This is good! Our bodies need to move solid waste out when we use the toilet. God gave us cellulose to help this happen. The bulkiness of the cellulose makes our intestines want to squeeze it along. It scrubs the intestines as it moves and carries other waste along with it. When talking about digestion, we call cellulose **fiber**.

There is more cellulose in the world than any other plant substance!

Time to do Activity 4 in the Activity Book!

Other Useful Things We Make from Plants

We are thankful for cellulose because of all the things it's possible to make and build with it. People have discovered how to make useful supplies from other ingredients found in plants:

Juice from the aloe plant helps heal burns.

1. Some **medicines** come from plants. Many plants are useful for first aid and other health problems.

2. Natural **dyes** that make cloth colorful are made from plants and minerals. To make dye, colorful plants or minerals are mixed with plant starch and seaweed which keep the color from washing out of the cloth.

3. The milky sap of **rubber** trees is collected to make forty thousand different products like rubber bands, medical supplies, toys, and tires. Tires are a mixture of manmade rubber and natural rubber. God's natural rubber helps keep the tires from cracking or tearing better than if they were made only of manmade rubber.

4. Some **paints** are made of plant oils and colors. Turpentine is made from pine trees and is used to thin oil paint, help paint dry faster, and clean paint brushes.

5. Fuel is something that is burned so its energy can be used. Coal, oil, and natural gas are **fossil fuels** that come from deep in the earth. We burn them to heat our homes and make our cars move. They are called fossil fuels because, like fossils, they are the remains of plants and animals that lived long ago. Before Noah's flood,

Mining Coal

large forests grew on the earth. These forests were torn up during the flood and then buried by sand. In time, they became coal. Some of this coal got buried even deeper underground. Its temperature became so hot that oil and natural gas were formed. This has happened in many places around the world.[1]

6. **Ethanol** is a fuel made from corn, barley, sugar beets, and sugar cane. Even though these are food plants that can give us quick energy, people have invented a way to change these plants' sugars into ethanol. The ethanol is then mixed with fossil fuel, delivered to gas stations, and we pump it into our cars.

7. These same quick-energy plants can also be used to make **plastic**. But most plastic is made from fossil fuels.

8. **Ink** is made from plant materials. The black ink for computer printers is made by burning fossil fuels and collecting the fine black powder left behind. The powder is then mixed into linseed or soybean oil from plants.

 Time to do Activity 5 in the Activity Book!

Prayer

 We praise You, Lord, for providing plants for supplies all over the earth. You have made cellulose and other plant products so useful to us. Thank You for giving us brains to invent ways to use the supplies in plants. Amen.

God has made wonderful gifts for us by creating plants! There are so many things that can be made from plant products.

Today is Green Thumb Day! Time to do Activity 6 in the Activity Book.

Cotton field

CHAPTER 3

God Fills the Earth with Busy Workers

When God created plants, He created many different ways that plants can give food and supplies to the whole world. But besides *giving* us things, plants *do* things! They are busy doing all the jobs God created them to do.

Let's learn a little about chemistry to help us understand photosynthesis!

Plants Work So We Can Breathe

Plants are busy all day long doing **photosynthesis**. Remember, photosynthesis is the way that plants use God's sunlight, water, and air and turn it into food for themselves and for people and animals. Let's learn another great thing that happens during photosynthesis!

Chemistry is studying the stuff things are made of. Everything is made of chemicals. Chemicals are made of **molecules** that are too tiny to see. A molecule is the smallest piece of something that still is that thing. If you hold a drop of saltwater on your finger, all you can see is water. But inside that drop of water are salt molecules and water molecules. If you could divide that drop of saltwater into smaller pieces, you could separate the water molecules from the salt molecules. Then you wouldn't have saltwater anymore. You'd have water, and you'd have salt. There is no such thing as a saltwater molecule. A salt molecule and a water molecule are very different things, but when they are mixed together, we get saltwater.

Molecules are made of smaller things called **atoms**. Let's pretend you divided your drop of saltwater into water molecules and salt molecules. Now pretend that you looked very closely at the water molecules. If you could see them, you'd see that a water molecule is made of three atoms joined together. Two of the atoms are **hydrogen** atoms and one is an **oxygen** atom. If these three atoms are separated, they're not water anymore.

A water molecule is called "H2O"

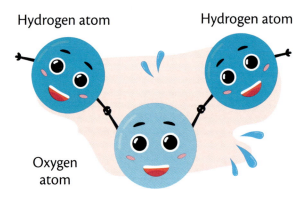

Hydrogen atom

Hydrogen atom

Oxygen atom

Sunlight is pure energy. It's not made of anything we can touch. It doesn't take up space. It's not made of chemicals. But sunlight is very important for plants. Plants need the light from the sun to make their food.

Plants use their roots to carry water up from the ground and into the rest of the plant. When a plant makes food, it uses the energy from sunlight to break apart the water molecules that have come up from its roots. The plant then uses the two hydrogen atoms from each water molecule as ingredients to make food (sugar). Plants don't need the oxygen atom that's left over, so the oxygen atom joins with another oxygen atom from another broken water molecule. Together they leave the plant and go into the air.

Air looks like empty space. It looks like it's made of nothing, but it is actually made of molecules too. Air looks different than water because the molecules in air are spread out far from each other. When molecules are spread out far and floating around, it's called a **gas**. The pairs of oxygen atoms leaving the plant are **oxygen gas**.

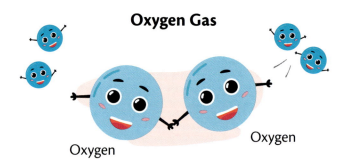

Oxygen Gas

Oxygen

Oxygen

This oxygen gas is what people and animals need to breathe! Plants make this oxygen for us. God has provided oxygen makers all over the earth for His creatures. He made plants able to busily change water molecules into the oxygen gas we must have to live!

Time to do Activity 7 in the Activity Book!

Plants Work to Improve Soil and Air

Good soil is important! Plants grow better and produce more food when they live in good soil. Plants busily help make soil better in two ways:

1. Plant roots put sugars and other good things into the soil to feed tiny creatures that are too small to see. These creatures then break up dead plants and minerals in the soil into things the plant uses. When different living things help each other in this way, it's called **symbiosis**.

2. Plant roots grow deep into the soil and hold the ground firmly in place. This helps prevent soil **erosion**. Erosion is when parts of something are gradually taken away and moved somewhere else. Soil can be eroded by being blown away by the wind or washed away by water. Plants help prevent erosion from happening.

Moist air is important. Plants take moisture from the soil and put it into the air. This moisture is extra water that is not needed for photosynthesis. Moist air is less drying to our skin and more comfortable to breathe. It makes us cooler too! In the summer, a large maple tree can put 50 gallons (190 L) of water into the air every hour. Having grass, trees, and shrubs in your yard can make the air temperature around your home up to 14°F (8°C) cooler.[2] This water from plants can also join with other water in the air and become clouds. The clouds rain down water for the plants to use again.

Plants on hills prevent erosion.

 Time to do Activity 8 in the Activity Book!

Prayer

Thank You for all the jobs that busy plants do, Lord! Using plants, You give us oxygen to breathe. You wisely gave plants a way to care for the soil they are growing in. And You make us comfortable with cool moisture from the plants we live with. Amen.

CHAPTER 4
Special Plants in Special Places:
Grasslands

A biome (bye-ome) is a large area in the world that has certain kinds of plants, animals, and temperatures. Let's learn about the grassland biome!

About Grasslands

Grasslands are large, mostly flat areas of grass. These areas usually have long, warm summers and short, cold winters.

God sends rain and snow on grasslands, but He only sends enough moisture for grasses to grow. There isn't enough water to grow a forest, but a few trees sometimes grow in grasslands.

Fires start more often in grasslands than in any other biome. Every winter, the grass freezes, dies, and gets flattened to the ground. After a few years, a layer

Prairie fires burn off dead grass.

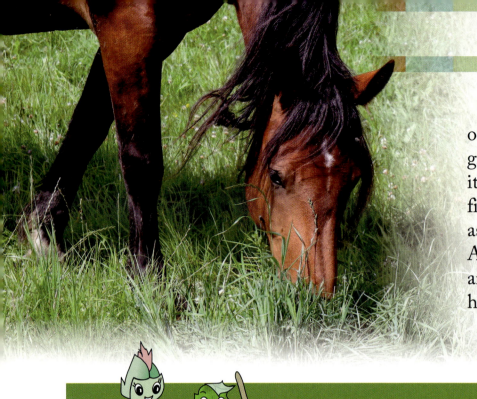

of dry plants has built up on the ground. When lightning strikes, it can start fires in this fuel. The fire quickly spreads and burns as the wind carries it along. Although fire can ruin things and be dangerous to life, it is helpful for grasslands.

Herby & Flora's List of Fire's Benefits

• Fire burns away dead plants, making it easier for new grass to reach the sunlight.

• Dead plants trap winter's chill in the ground and keep it there so new grass can't grow when spring starts. Fire gets rid of this layer of dead plants. Once these dead plants are burned off, new grass can start growing earlier in the spring.

• Fire takes the nutrition from dead grass and puts it into the soil for the new grass to use. If fire didn't do this, the nutrition would stay in the dead plants a very long time, and new plants wouldn't be able to use it.

• Grass roots can survive a fire better than most other plants, shrubs, or trees. Fire keeps grasslands from changing into other biomes because it kills the bigger plants.

• With dead plants burned away, grazing animals have an easier time finding good grass to eat. And the new grass that grows is healthier! The animals quickly grow fat because they don't have to use up their energy trying to find good green grass.

Sometimes people purposely start fires on their land to make grass better for grazing animals.

Fire does good things for the grasslands! Maybe that's how fire praises the Lord as it's told to do in this psalm.

Praise the LORD from the earth,
You great sea creatures and all the
depths;
Fire and hail, snow and clouds;
Stormy wind, fulfilling His word.
(Psalm 148:7-8)

Time to do Activity 10 in the Activity Book!

Special Plants in Grasslands

We have already discovered the amazing way God provides food to people and animals all over the world with fruitful grass. Using grass, He provides grain for people and pastures for grazing animals. Let's look at grass a little more closely.

God has given many grasses ways to make more plants even if their seeds are somehow destroyed. Some special grasses have underground stems which keep growing sideways secretly. They will make roots every now and then, and a new grass plant will spring up from these roots. Then the secret stems spread underground to do it again. Other special grasses have stems that creep sideways on top of the ground and grow more grass plants at their ends. The new plants send down roots and can make new sideways stems of their own on the surface of the ground.

Grasses are also wonderfully fruitful because grass **blades** (leaves) can grow back easily after an animal eats them. Most other plants grow taller only from their tips. If you cut the top off a tree, it won't get any taller. The reason for this is that plants have certain places where they grow. The growing places that make trees taller are only at the tips of the branches. Grasses are amazing because God also gave them growing places down near their **crowns** (where they come out of the ground). This means that, even though the tips of the blades might be eaten off, the blades will still get taller by growing from the bottom.

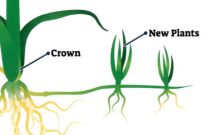

Parts of a Plant

Seedhead

Blade

Crown

New Plants

We should be happy when God shows us things we need to change. When people obey God, He says they can be as fruitful as grass!

"Happy is the man whom God corrects; Your descendants shall be many, And your offspring like the grass of the earth." (Job 5:17,25)

Since grasslands have very few trees to supply wood, settlers in these areas often made homes by stacking pieces of sod to make walls.

This roll of grass and roots came from a sod farm. It will make a nice lawn if kept watered and mowed.

A flood in this creek washed away soil and allows us to see the tangled roots of grass.

Time to do Activity 11 in the Activity Book!

Grasslands have other short plants besides grasses. Purple coneflowers are one of the hundreds of kinds of plants in grasslands.

Saiga antelope of the steppes

Animal Life in Grasslands

Some of the animals living in the grassland biome are: bison, wild horses, pronghorn, antelope, prairie dogs, coyotes, falcons, snakes, birds of prey, and different kinds of insects and gophers.

Heavy grazing animals are important for the grasslands. God made these animals wander while they eat. Here are some reasons why grazing animals are perfect for keeping the grass healthy:

• When grass is partly eaten or cut, it sends itself a signal to make more stems that travel sideways underground or along its surface. This makes even more grass in the grassland than would grow from seeds!

• Tall grass makes too much shade. Grass that grows in its own shade isn't as strong or healthy as other grass. Grass roots don't grow well when the ground above them is shaded. Wandering animals help by eating the tall parts of grass. Then the sunlight can shine on the grass underneath.

Prairie chicken

- Sometimes grasslands don't get much rain. When this happens, grasses might lose too much water from their blades. When grazing animals keep the grass blades short, the plants lose less water.

- With their sharp hooves, grazing animals scratch away some of the dead plants on the ground while they walk and graze. Then they move along to a new spot. This makes little areas of bare ground where seeds can fall and touch the soil and sprout into new grass.

- When the ground is soft after a rain, heavy animals can make deep

Przewalski's horse of the steppes

footprints in it. This is good because each footprint can collect water like a little pond the next time it rains. The water in the footprint does not drain away or evaporate quickly. It slowly soaks into the soil and keeps plant roots and tiny soil creatures alive. Wandering animals make lots of footprints!

- Animal manure fertilizes the grassland plants.

Most of the grazing animals on Earth belong to people because people use their products. This means we can't let

Definitions

The **prairies** are the grasslands of North America.

The **steppes** are the grasslands of Europe and Asia.

Bison of the prairie

Prairie falcon

should know when to move the animals to another pasture to keep them from eating so much that the grass can't grow back. He needs to make sure the heavy animals aren't in one place so long that they make the soil hard. He is moving his animals to help his land the way wandering animals would. He is taking care of his land as God told Adam to do in this verse:

Then the LORD God took the man and put him in the garden of Eden to tend and keep it. (Genesis 2:15)

them wander like wild animals. We have to either feed them or let them graze in fenced pastures.

When useful animals are grazed, the owner needs to know his pastures' grasses. He should know how much grazing will help each pasture. He

Food products from animals that eat grass are healthier for us than products from grain-fed animals. When an animal grazes, it eats many different kinds of grass and other plants in the grassland. Its body makes the kind of fat that's good for us to eat. When an animal is fed only a few of the kinds of food God made it to eat (like a lot of grain), its body makes a different kind of fat that isn't healthy for us.

Different grasslands around the world have different large animals living on them. The grasslands of North America are called **prairies**. At one time, no one owned the prairie land. Millions of wild bison lived on the prairies and kept the grasslands healthy.

Then people brought cattle to the

Prairie rattlesnake

prairies. At that time, the cattle in an area belonged to different owners, but they all grazed together, keeping the grassland healthy. Each cow was branded with a mark that told who its owner was. When the cattle were large enough to be sold for meat, cowboys working for each owner would sort out his boss's cattle and take them to a railroad. On these long trips across the grasslands, there might be ten cowboys in charge of 3,000 cattle. These **cattle drives** could take up to two months. They had to go slowly and allow the cattle to graze twice a day so they would be fat enough for someone to want to buy them.

Now, because people own the land and put fences on it, there is no longer a large path of wild grassland to drive cattle on. Highways make it easy to send cattle on trucks to where they are sold. People who own grazing cattle today

Prairie dogs

keep them on land called **ranches**. These "ranchers" need to move their cattle around to different fenced pastures to best take care of their grassland ranches.

The large grasslands of Europe and Asia are called **steppes**. There were once many wild horses and antelope grazing there. As the area became more settled, cattle and sheep became the main grazing animals.

Time to do Activity 12 in the Activity Book!

Prayer

Grass is wonderful, Lord! It can be fruitful in so many ways: seeds, sideways stems, and blades that grow from the bottom. It provides grain for us and food for animals. And You even use dangerous fire to help keep grassland fruitful. Thank You for grasslands! Amen.

UNIT 2
Flowers: Beautiful Crowns

Flowers have a job to do even though they are pretty. God gave them the important job of making seeds!

I'm so happy to be a flower! I loved it when God made me slowly bloom and quietly wait until an insect came to visit. And when glad children came to see my colors and smell the sweetness, I wanted to twirl around on my stem!

 ## Memory Verse

"The grass withers, the flower fades, But the word of our God stands forever." (Isaiah 40:8)

 ## Hymn to Sing

You can listen to this hymn by searching for "God Sees the Little Sparrow Fall, children's hymn" on the internet.

God Sees the Little Sparrow Fall

God sees the little sparrow fall,
It meets His tender view;
If God so loves the little birds,
I know He loves me too.

Chorus:
He loves me too, He loves me too,
I know He loves me too;
Because He loves the little things,
I know He loves me too.

He paints the lily of the field,
Perfumes each lily bell;
If He so loves the little flowers,
I know He loves me well.
(Chorus)

God made the little birds and flow'rs,
And all things large and small;
He'll not forget His little ones,
I know He loves them all.
(Chorus)

Most flowers open very slowly. But the flowers of the baobab tree open quickly enough for people to see their petals move!

CHAPTER 5

God Amazes Us with Flowers

God Amazes Our Eyes

Do you like to look at green fields, forests, and gardens? Green is a restful color for our eyes. But it's also a beautiful background for the surprise colors of flowers! People can tell the difference between 10 million different colors.[3] But there are some colors that we wouldn't see outdoors if it weren't for flowers. In fact, our names for some colors (pink, violet, and fuchsia) come from flower names.

Each kind of flower has its own special shape. Flowers can be simple or complicated. They might be all alone on a plant or be smashed together with more than a thousand just like them. Some turn their faces to the sun. Others look shyly toward the ground. The world is full of an amazing variety of flowers!

"Consider the lilies, how they grow: they neither toil nor spin; and yet I say to you, even Solomon in all his glory was not arrayed like one of these." (Luke 12:27)

Jesus talks about the beauty of lilies in this verse. There are about 90 different kinds of lilies! They come in all colors except blue. Some smell luscious, and others have no scent at all.

Mariposa lily

Fuchsias

light in the late afternoon. Inside the plant's leaves there is a certain chemical that checks on the amount of light there is each day in the late afternoon. This chemical senses when the days are still too short and knows it isn't time to do its job. As the days keep getting longer, the sun shines brightly later and later

How do plants know when to make flowers? Their leaves tell them! Plants begin to grow in the spring. They need sunlight to help them grow. Winter days are short, but in spring the days start to get longer. The sun stays up later than it did in the winter, but there still isn't much

Violets

Sometimes the same kind of flower comes in many beautiful colors. Colorful tulips are raised in Holland.

The bird of paradise flower looks complicated.

The passion flower looks too strange to be real!

into each afternoon. One day, when the special chemical does its late afternoon check, it finds that the time has come to do its job!

The chemical travels out of the leaves and up to the tips of the plant where the special growing places are. Up until now, the growing places have been making the plant grow larger. But now the chemical tells each growing place to change its job. It tells it to start making flowers! This is an amazing way that God made plants able to know the right time of year to make a change.

Sometimes, if the weather is still too cold in spring, flowers need to wait to **bloom**. If flowers open too early, they can freeze and die. Then they can't do their seed-making job. And if the weather hasn't warmed up yet, insects won't want to come out of their winter homes to help flowers make seeds. To keep flowers from blooming

too soon, God made God made plants' roots and parts above ground to know when the temperature is too cold. The plant is able to tell the flowers to wait!

God has another way to make sure that flowers don't bloom too early in the year. Sometimes, there are short warm spells in the middle of winter, but there will be several more months of freezing temperatures to come. If warm weather was the only signal for flowers to bloom, they could be fooled into blooming at the wrong time. To keep this from

The amount of afternoonsunlightislike a calendar telling the plant that it's a special day: a special day to work!

53

All the beautiful apple trees in an orchard bloom at the same time. God has given the same chemical to all these trees so they know when spring has come and they can start making flowers.

Young sunflowers turn to follow the sun all day long. Their centers are made up of crowded **disk flowers**. Yellow **ray flowers** line the circle's edge.

happening, God taught some plants that they must go through a certain number of cold winter days before they bloom. The plants count the number of cold days to make sure there have been enough of them to equal a whole winter.

Have you noticed that some flowers bloom in early spring and others wait until summer or fall? This is important because bees and other insects need flower **nectar** to eat in spring, summer, and fall.

Peach trees, hollyhocks, and daffodils are plants that need to count cold days. They wouldn't bloom if we planted them in warm parts of the world because there aren't enough cold days

Butterfly on fall chrysanthemums

Definition

When a plant **blooms**, its flowers open.

We learned that flowers need long enough days before they can bloom. Spring flowers can start blooming when the days are still quite short. But when the days get longer in summer, they stop blooming! By then the summer flowers have started blooming because the days have become long enough. But there are some plants, like tomatoes, that don't seem to mind how much light they get. They just bloom when they get old enough!

God always provides something for insects to eat. And He amazes our eyes with beautiful flowers in the spring, summer, and fall.

Tomato plants bloom when they reach a certain age.

Time to do Activity 13 in the Activity Book!

God Amazes Our Noses

God has given us special places in our noses that sense smells and send the information to our brains. Our noses help us know when there is a danger like fire or spoiled food we shouldn't eat.

But God is so loving to make our noses also able to sense nice smells like warm cookies, fresh laundry, and flowers. Part of the beauty of flowers is their **fragrance**.

A flower's smell is created in the petals where different chemicals are.

The three-foot wide (1 m) meat flower attracts pollinating flies with a smell like rotten meat.

Blackberry flower and its daytime pollinator

The molecules of these chemicals don't weigh very much, so they easily float into the air, mix with each other, and make **scents**. Flowers pollinated by bees and flies usually have sweet **fragrances**. Flowers pollinated by beetles have strong musty, spicy, or fruity scents.

Flowers need the help of **pollinators**. These pollinating insects or animals transfer pollen between flowers so that seeds can be made. Flowers attract pollinators with their scent! A flower will make the most scent at the time when its pollinator is active. Plants that are pollinated by bees or butterflies make

Definitions

A **pollinator** is an insect, bird, or bat that transfers pollen to the places in flowers where it's needed to make seeds. Pollination happens as the pollinator gets nectar and pollen as food for itself and maybe for its young.

A **fragrance** is a pleasant, sweet smell.

56

If you want to see the face of the shy maiden flower, or smell its fragrance, you must lie on the ground to get near it. It's less than five inches (12.5cm) tall and won't look up!

the most fragrance during the day when these insects are active. Plants that are pollinated by nighttime moths or bats put out the most fragrance at night.

Flowers also make the most scent at the exact time they are ready for a pollinator's help. Flowers that are too young or too old to make seeds have less fragrance. The pollinators would rather go to a fragrant flower. God is the wise One who made flowers and pollinators this way so they can work together! The flower feeds the pollinator, and the pollinator helps the flower make seeds.

You open Your hand And satisfy the desire of every living thing. (Psalm 145:16)

This verse reminds me of when flowers open and offer their pollen and nectar to their pollinators. Only God can open His hand to feed all the people and creatures of the world!

Time to do Activity 14 in the Activity Book!

A bat gets nectar and pollinates at night.

Prayer

Lord, You are so wise and amazing to make colorful, fragrant flowers! You made them to know just when to form on the plant and when to open. You give them scents so they can attract the right pollinators for seed making. Thank You for showing us Your wisdom when we learn about Your lovely flowers! Amen.

The gardenia's sweet, creamy fragrance gets stronger at night to attract moths.

Flora is hiding somewhere nearby. Can you find her?

Today is Green Thumb Day! Time to do Activity 15 in the Activity Book.

CHAPTER 6
Parts of a Flower

For lo, the winter is past,
The flowers appear on the earth;
The time of singing has come.
(Song of Solomon 2:11-12)

When flowers appear, they come in many different colors, shapes, and sizes. But flowers all have the same purpose: to make seeds that will grow into more plants. Because of this, they all have certain parts. Let's learn what these parts are so we can understand seed making!

Basic Flower Parts

The picture on the next page shows the parts of a flower. These parts may look different in different kinds of flowers. And some flowers may not have every part. Let's learn the basic flower parts!

Sepals are the green leaves just below the flower. Before a flower opens, it's called a **bud**. When a flower is a bud, the sepals are wrapped around it to protect it.

Petals are the colorful, fragrant parts of flowers. Petals attract pollinators by their colors and smells. Bees are pollinators that can see **ultraviolet**, a color that's invisible to us. Flowers that need bees for pollination often have ultraviolet markings that lead bees to the center of the flower. There, the petals have made sweet nectar for them to eat. It's wonderful that God

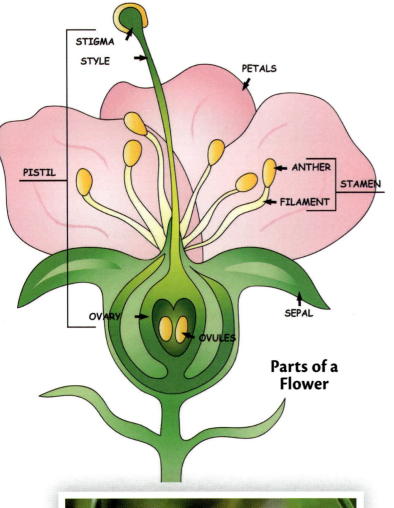

Parts of a Flower

A butterfly pollinates flowers as it drinks nectar.

has made symbiosis between flowers and pollinators! The flowers feed their pollinators, and the pollinators trade pollen between flowers so the plant can make seeds. Without this symbiosis, the world would be missing many plants and foods.

Stamens are the pollen-making parts of a flower. They are made up of the anther and the filament.

Anthers are the pollen-making machines. They lightly hold the pollen until it's taken away by a pollinator or by the wind.

Filaments are the stalks that hold the anthers out so the pollen can be carried away.

Pistils are the pollen-taking, seed-making parts of flowers. They are made up of the stigma, style, and ovary.

The **stigma** is a sticky place whose job it is to collect pollen.

The **style** is a stalk that holds the stigma up where it can reach pollen that comes by.

The **ovary** is the special place that holds the ovules. The ovary becomes the fruit of the plant.

The **ovules**, or eggs, become the seeds of the plant.

 Time to do Activity 16 in the Activity Book!

Flowers with Missing Parts

Some plants don't need bird, bat, or insect pollinators. They are pollinated by the wind! Their pollen is so light-weight that the wind can easily carry it. God has given wind-pollinated plants a lot of extra pollen so that there is plenty blowing around to be caught for seed making. Wind-pollinated plants, like grasses and many trees, have very small petals or no petals at all on their flowers because they don't need to attract pollinators.

Pine trees are pollinated by the wind. Pollen can make the air look dusty when the trees let it go.

Pollen-making flower

Pollen-taking flower

Squash and pumpkin plants have large, colorful petals, but each flower is missing either the pollen-making or the pollen-taking parts. On the same squash plant there will be some flowers that have the stamens to make pollen and other flowers that have the pistils to take pollen. They are pollinated by insects.

Hezelnut catkins make pollen

Pollen-taking flowers

Each hazelnut tree has separate pollen-making and pollen-taking flowers. **Catkins** are long, hanging groups of pollen-making flowers. They have only stamens. Wind blows their pollen to the pollen-taking flowers. These pollen-taking flowers have only red stigmas showing.

Corn tassels make pollen.

Corn silk takes pollen.

Corn tassels, high on the plant, have stamens for making pollen. Wind takes the pollen down to the corn silk where the ears of corn will be. Each string of corn silk is the stigma and style of one kernel of corn.

Pollen-making holly flowers

Pollen-taking, berry-making flowers

Some plants, like holly, asparagus, dates, and spinach, have their pollen-making and pollen-taking flowers, not just on separate flowers, but on separate plants! If you want to grow holly berries, you will need to have two separate bushes: one with pollen-making flowers, and the other with pollen-taking flowers.

Prayer

Oh Lord, whenever we see and study the complicated beauty of flowers and all their parts, help us remember You. We want to praise You because You made so many kinds of flowers for us to enjoy. Amen.

Some plants, like tulips, look like they are missing sepals. But they aren't! Tulips have three sepals that look just like their three petals.

Time to do Activity 17 in the Activity Book!

Isn't God beautiful? He created so many lovely flowers and gave us eyes to see them!

Do you know what kind of flower I am? I'm a hollyhock!

Today is Green Thumb Day! Time to do Activity 18 in the Activity Book.

A hummingbird
drinks nectar
and pollinates.

CHAPTER 7
Seeds Begin in Flowers

Most plants make seeds. Seeds grow into new plants that will also make seeds. God made seeds this way so there will always be plants on Earth.

Then God said, "Let the earth bring forth grass, the herb that yields seed, and the fruit tree that yields fruit according to its kind, whose seed is in itself, on the earth"; and it was so. (Genesis 1:11)

Seed Making

Most of the plants on Earth make flowers. When these flowers are pollinated, the process of seed making has begun. Let's learn from Herby how flowering plants begin to make seeds.

Seed making takes a lot of work, but I like work! Here's the process:

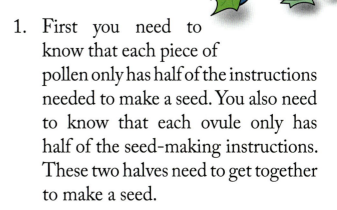

1. First you need to know that each piece of pollen only has half of the instructions needed to make a seed. You also need to know that each ovule only has half of the seed-making instructions. These two halves need to get together to make a seed.

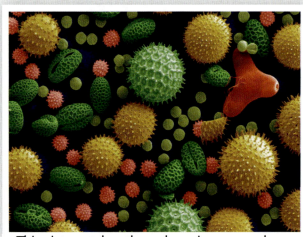

This picture, taken through a microscope, shows pollen from several kinds of flowers.

2. Pollinators, such as hard-working bees, visit flowers. God created bees to visit only one kind of flower on each trip from the hive. This is not important for the bees, but it is important for the flowers. Flowers cannot make seeds with the pollen from a different kind of flower. They need their own kind of pollen to make seeds. Pollinators don't know that God has given them this job of giving the right kind of pollen to the right flowers. It's just something they do as they work at getting food for themselves. God has wisely made bees to help plants make seeds according to their kinds!

3. As a pollinator crawls around on the flower, it's trying to find sweet nectar or nutritious oils. It may also be collecting some of the pollen for protein. Pollen is sticky, and pollinators are fuzzy. Even if the pollinator isn't trying to collect pollen to eat, some of it easily sticks to its fuzz when it brushes up against the flower's anthers.

This honeybee will carry nutritious pollen back to its hive in the "pollen baskets" on its back legs.

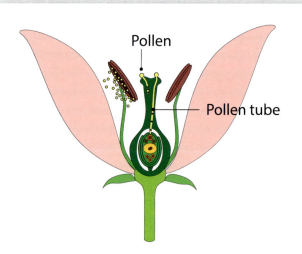

Pollen

Pollen tube

Half of the seed-making instructions travel down the pollen tube to an ovule. The ovule contains the other half of the seed-making instructions. Once they join together, the ovule has all the necessary instructions to become a seed!

4. When the pollinator can't find any more food in one flower, it moves to another flower and brings along the pollen from the first flower on its fuzz.

5. As the pollinator explores the next flower, some of the pollen from its body gets stuck on the sticky stigma of the new flower. This is just where God wants it to be because now seeds can be made!

6. Pollen comes in different shapes for different flowers. Only pollen of the right shape will be allowed to give its half of the seed-making instructions to the flower. If the right piece of pollen gets stuck on the stigma, it grows a tiny tube that travels all the way down the style and into the ovary.

Pollination

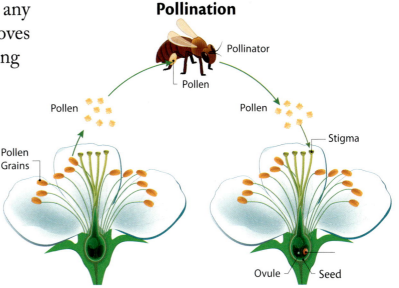

7. This tube is the passageway for the pollen's half of the seed-making instructions to travel down. At the bottom, the pollen's instructions join with the ovule's half of the instructions.

8. Now, with complete instructions, the ovule can grow into a seed!

Time to do Activity 19 in the Activity Book!

Flowers and Pollinators Get Together

Many pollinators, like honeybees, can pollinate many kinds of flowers like dandelions, roses, and apple blossoms. And many flowers can be pollinated by several different kinds of pollinators, like bees and flies.

But God made some flowers and pollinators that can only work with each other and nothing else. The yucca moth and the yucca plant work together.

A pollen wasp pollinates flowers while it eats nectar and pollen.

We have learned how flowers attract pollinators by using their colors and fragrances to promise delicious nectar. Let's look at some other amazing ways God made for pollination to happen!

Bird Beaks and Flower Shapes That Match

Without the yucca plant, there would be no yucca moths because the baby moths need to eat some of the plant's seeds to survive. Without the yucca moth, there would be no yucca plants because the moths are the only insects that can pollinate yucca flowers.

Flowers that are pollinated by hummingbirds often have long tubes with nectar at the bottom. The hummingbirds must stick their long beaks deep into the tubes to reach the nectar. When they do, their face feathers pick up pollen from the flower's stamens. That pollen gets carried to the next flower they visit and rubs off onto the flower's stigma.

Blooming yucca plant and yucca moth

Puff Pollination

One kind of flower has a spongy pouch connected to each stamen. This pouch is full of sweet nectar. When a bird tries to squeeze nectar out of one of these pouches, he also squeezes a puff of pollen into his face. Then he carries the pollen to the next flower on his face feathers!

A Take-Along Scent for Bees

Bucket orchids make a scent that a certain green bee likes. The bees like to rub the scent all over themselves. But, in their excitement, they often fall into the flower's bucket of water. The only way out of the slippery bucket is through a narrow tube that has helpful

bee-sized steps. As the bees climb up these steps, they squeeze past the stamens and pick up pollen to bring out with them.

Fig flowers shown in an open fig and fig wasp.

Bucket orchids and their green bee pollinators

Flowers & Wasps Hide Together

There are about 800 kinds of fig plants, and each one has its own kind of wasp for pollination. Figs have very interesting flowers. If you hold a fig in your hand, you are not holding a fruit. You are holding a special stem that has become a bulge. The inside of the bulge was once lined with hundreds of little flowers. It was dark inside. Only a tiny hole was open to the outside. The opening was so small that only a certain kind of female wasp could squeeze in to pollinate the flowers, and she even lost her wings and antennae while squeezing through. But she wanted to go in to lay her eggs. She brought along pollen from the fig where she had hatched and grown up. The hidden fig flowers were pollinated as she laid her eggs.

When the mother wasp's eggs hatched and grew up into adult wasps, they chewed another hole through the fig and flew away. The female ones carried pollen to another fig with hidden flowers. Pollinated fig flowers become hidden fruits inside the stem bulge. When you chew a fig, you can feel the tiny seeds from those hidden fruits in your teeth.

Buzz Pollination

Some plants, like blueberries and tomatoes, can be partly pollinated by wind or by honeybees. But this pollination doesn't work well because the flowers have very small openings for the pollen to go through. God has helped these plants by giving them bumblebees who can make their flowers release four times more pollen. Tomato flowers don't make nectar, but bumblebees love their pollen. Bumblebees and a few other bees are able to make certain flowers vibrate by buzzing! When they vibrate, the flowers release a lot more pollen out of their small openings. Honeybees can't do this. Their buzzing is not the right sound. But bumblebees can make just the right sound to vibrate the flowers and make them give the greatest amount of pollen.

Bumblebee
pollinating a
tomato flower

Time to do Activity 20 in the Activity Book!

Prayer

Dear God, how amazing You are to have created perfect ways for pollen's instructions to get to the ovules. You have lots of pollinator helpers that cheerfully, and unknowingly, deliver pollen while they are rewarded with food. And then the pollen builds a safe passageway to send its instructions through to the flower's special seed-making place. How wonderful that these things are happening wherever there are flowers! Amen.

CHAPTER 8

Special Plants in Special Places:
Deserts

About Deserts

Deserts are dry! The desert biome gets much less rain than the grassland biome. South America's Atacama Desert, the driest place on Earth, once went 14 years without rain! When rain does fall in the desert, it might be a small amount a few times a year. Or it could come only once a year but be a flooding downpour. Either way, it still is very little rain.

Deserts are very hot in the daytime. They are almost always sunny because there are almost never clouds in the sky to shade the ground. Also, there isn't enough moisture in the dry desert air to help keep the daytime cool.

Deserts are cold at night. If there were clouds in the sky or moisture in the air, warmth would be held near the ground. Without clouds, the day's heat quickly goes away, and the land becomes

Rainwater quickly drains through the rocks, sand, and gravel of desert soil.

cold.

Desert soil is made of sand and gravel. Water quickly drains through it. But desert soil has a surprising amount of nutrition. If water can be brought in, crops will grow very well!

Time to do Activity 22 in the Activity Book!

Desert plants are spaced far apart.

Special Plants in the Desert

Grasslands seem so full of living things, but deserts seem so empty. In the desert, we usually see very few plants and a lot of bare ground. Let's learn about desert plants!

There are two main types of plants in the world: **annuals** and **perennials**.

O God, You are my God;
Early will I seek You;
My soul thirsts for You;
In a dry and thirsty land
Where there is no water.
(Psalm 63:1)

King David wrote this psalm when he was hiding from danger in the desert! Even in the desert, God protects His people.

Definitions

Annual means something that happens once a year. Plants that grow only from seeds are called **annual plants**, or **annuals**, because they sprout, grow, bloom, make seeds, and die all in one year.

Perennial plants, or **perennials**, have roots that don't die even if the rest of the plant freezes or dries up. The stems, flowers, and leaves above ground may disappear, but new ones grow from the living roots when the weather is just right. Bushes and trees are perennials whose stems don't die.

Perennial Desert Plants

The plants we see spaced far apart in a desert are perennials that don't die during the desert's dry times. They are spaced far apart because there isn't enough water for crowded plants.

How are these few desert perennials able to stay alive in such hot, dry places?

1. God made perennial plants of the desert able to collect water so they can live and get bigger each year:

   • The saguaro (suh-wah-roh) cactus has long, sideways-spreading roots that don't grow very deep in the ground. These roots are able to collect rainwater as soon as it falls and before it dries up.

   • Desert trees live in low places where water may run and sink in after a rain. These trees have long

Mesquite (mess-keet) trees have roots that reach as deep as 100 feet (30 m) under the ground.

roots to collect water that has gone deep underground.

   • There is a desert moss that collects water from the air during the few times there is fog or rain in the desert. Each tiny leaf tip ends in a hair that sticks up and catches water. When there is enough water on the hair, a drop forms and slides down to the leaf. The

Saguaro cacti can grow 40 feet (12 m) tall. Their roots would then stretch 40 feet from the trunk in all directions.

Tortula moss has moisture-collecting hairs at the end of each leaf.

leaf opens up, takes in the water, and quickly does photosynthesis using that water!

2. God has also given desert perennials ways to *hold onto the water* they get:

    -   The green parts of cacti are their stems. Cacti stems have two important jobs: to do photosynthesis and to hold on to water. The stems swell up with water when it rains. Then they shrivel as the cactus uses up its water in the dry times. Instead of leaves, cacti have spines which help prevent most thirsty animals from taking bites out of their water storage.

    -   Many desert plants have a waxy covering on their leaves or stems to help keep water inside.

Cacti have protective spines instead of leaves.

-   Water is lost through a plant's leaves. To hold on to water, some desert plants drop their leaves and go to sleep during the long dry times. After it rains, they sprout new leaves. Other plants die back to their water-storing roots when it's dry. They grow new stems and leaves when the rain comes.

-   Plants lose water into the air during photosynthesis. Many desert plants do photosynthesis a little differently. They separate the stages of photosynthesis so that the water-losing stage happens at night. Cooler nighttime temperatures mean less water will be lost!

Annual Desert Plants

If you explore a desert during its long dry time, you won't see any annual plants.

Can you tell which cactus is storing more water?

Agave plants are grown for their sweet sap. They do photosynthesis at night so they won't lose as much water.

Annuals avoid dry times by existing only as seeds when there is no rain in the desert. The seeds of desert annuals have thick, hard covers to protect their tiny living parts from drying up. God made these seed covers to open only when there has been at least 1/2 inch (1 cm) of rain. This way He makes sure there is enough water for plants to live and grow. When desert annuals grow, they can fill up the empty spaces between the perennials until the dry times come again.

Desert annuals make a lot of seeds. If you take a walk in the Coachella Valley of the California desert, you might be stepping on as many as 4,000 seeds each time you take a step! When a desert plant makes seeds, it makes some that will sprout the next year. It makes others

Sand verbena is an annual desert plant that sprouts, grows, and blooms after the winter rains.

that will wait and sprout in two or more years. This way, even if there is a dry year coming when seeds can't be made, God has saved some old seeds in the ground

79

to sprout. Desert soil might have seeds waiting ten years to grow!

Desert annuals have very short lives– maybe only 6-8 weeks. They sprout quickly from their seeds. Then they race to grow, bloom, and make more seeds before they dry up and die. These new seeds will sleep in the ground until everything is just right for them to sprout.

Desert Indian wheat races through life in just 2-3 weeks!

Desert flowers need insects to pollinate them. Where do the insects come from? They have been hiding as eggs or babies during the dry times. God made these insects in a special way. The same rain that causes the flowers to bloom causes the insects to grow into flying adults and come out of their desert hiding places!

The rain God sends to the desert makes the pollinators come out of hiding at the same time it makes the flowers bloom!

Time to do Activity 23 in the Activity Book!

Animal Life in the Desert

Black-backed jackal

"The beast of the field will honor Me,
The jackals and the ostriches,
Because I give waters in the wilderness
And rivers in the desert."
(Isaiah 43:20)

Besides jackals and ostriches, some of the animals that live in the desert biome are ants, scorpions, tarantulas, lizards, snakes, tortoises, toads, roadrunners, cactus wrens, gerbils, kangaroo rats, armadillos, jack rabbits, peccaries, camels, ibex, caracals, and kit foxes.

Like all animals, desert animals need water to live. Let's look at some desert animals and see how God gives them water in the dry desert.

Many animals, like the Gila woodpecker and desert cottontail rabbit below, get water from the plants they eat. Collared peccaries get water from roots they dig up.

Collared lizard

Tortoise

Cottontail

Gila woodpecker

Peccary

Honeypot ants

Honeypot ants live in cool underground burrows where the moist air keeps them from drying out quickly. Some of the ants make trips outside to collect nectar when flowers are blooming. Then they come back to the colony and feed this nectar to other ants that are hanging from the underground ceiling. The hanging ants store the nectar in their bodies for all the ants to drink during dry times.

Kangeroo rat

Camels

Kangaroo rats and camels can store fat in their bodies and turn it into water later. Each pound of fat in a camel's hump can turn into a little more than a pound of water! Camels and kangaroo rats don't pass urine very often. When they do, the urine is very thick because their bodies have saved the water.

Seeds are the main food in the desert. There are many rodents in the desert, and they all eat seeds. Some desert birds and insects also eat seeds. These seed eaters then become food for reptiles, tarantulas, foxes, and other meat eaters. Desert meat eaters can get half of the water they need from the juicy animals they eat.

Like all animals, desert animals would die if their bodies got too hot.

Almost all desert animals stay cooler by spending the hot part of the day in the shade or underground. Underground burrows are cooler than the sunny surface of the ground. Burrows also have moist air which helps animals save water. Early morning, evening, and nighttime is when desert creatures come out to eat,

A beetle provides food and water for tarantulas.

A roadrunner has caught a lizard for its food and water needs.

hunt, and meet each other.

The desert rodents below spend the day in burrows and come out in the cool nighttime.

Jerboa

Mongolian gerbil

Caracals, kit foxes, and jack rabbits have large ears that let off extra body heat as blood moves through them. This helps keep these animals cool even in hot weather.

Caracal

Kit fox

Jack rabbit

Desert air is very hot next to the ground. God gave ostriches and camels long legs to keep their bodies farther from the ground's heat. Vultures spend their days soaring high where the air is cool.

Just imagine how the animals of the desert rejoice when God sends rain!

Time to do Activity 24 in the Activity Book!

Prayer

Dear God, help us to be like King David when he was hiding in the desert: when we are hot and thirsty and looking for water, remind us that we should look for You in the same way. Thank You for filling deserts with seeds. Thank You for making special plants that can gather and hold water. We praise You for caring for desert animals and giving them water, food, and ways to stay cool! Amen.

Ostrich

UNIT 3
Beautiful, Busy Leaves

 ## Hymn to Sing

You can listen to this hymn by searching for "Come, Ye Thankful People, Come" on the internet.

Come, Ye Thankful People, Come

Come, ye thankful people, come,
Raise the song of harvest home;
All is safely gathered in,
'Ere the winter storms begin.
God our Maker doth provide
For our wants to be supplied;
Come to God's own temple, come,
Raise the song of harvest home.

All the food we harvest from plants is only possible because of the work done by busy, beautiful leaves!

 ## Memory Verse

"God did good, [and] gave us rain from heaven and fruitful seasons, filling our hearts with food and gladness." (Acts 14:17)

CHAPTER 9

Lovely Leaves

How Leaves Look

God made leaves in many beautiful shapes, with lovely colors and interesting edges. Some leaves are shiny. Some are thick and juicy. Some are so big you could shade yourself with just one. Other leaves are so small that they make their tree look misty.

Warbler in palo verde tree

The palo verde tree looks blurry because of its tiny leaves.

Succulent plants come in many colors. Many have thick, juicy leaves.

leaf parts! Knowing these parts will help us when we talk about plants.

The **blade** is the main part of a leaf. Photosynthesis happens mostly in leaf blades. Aspen leaves have round blades. Grass blades are long and thin. The shape of the blade helps us identify the plant.

Veins take water and minerals that have come from the roots and deliver them to the leaf blades. They also collect food made by the leaf and move it along so it can be used by other parts of the plant.

It's exciting to watch leaves move when they are blown by the wind. In grasslands, wind makes the tall grasses sway up and down like ocean waves. Isn't it fun to kick up fallen leaves on a windy autumn day and watch them blow away?

Definitions

To **identify** something means that we know what it is.

The way leaves look can help us figure out what kind of plants they are growing on. Certain books, called field guides, show the shapes of different leaves to help us **identify** plants. Let's learn some

Parts of a Leaf

Blade

Veins

Petiole

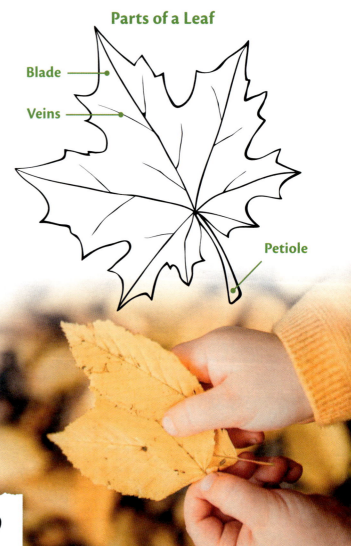

The arrangement of leaf veins helps us with identification.

Veins in the leaves of skunk cabbage

A **petiole** is a thin stalk that attaches a leaf to a stem. But most long, thin leaves, like grasses, don't have petioles. Their blades and veins attach directly to the stem.

Sometimes it's hard to tell the difference between a leaf and a leaflet. But you can tell by looking at the place where the petiole joins the plant stem. If there is a bud forming for next year's leaf, then you are looking at a leaf. If a bud isn't forming there, you are probably seeing a leaflet. A bud never forms at the bottom of a leaflet.

The ash tree below has leaflets. Notice there are no buds below each leaflet.

Even in a light breeze, the trees in an aspen forest look like they're quaking. This happens because the leaves are attached to the twigs by long, flat petioles that allow them to twist easily.

Leaflet

Petiole

Bud

Some leaves are divided into smaller **leaflets** that look like leaves themselves.

Leaves are specially arranged on their plants so that they don't usually touch each other. This helps the right amount of sunlight shine on each leaf. But what sound do you hear when the wind blows through leafy trees? Do you hear the leaves hitting each other? The trees are doing what this verse says!

"And all the trees of the field shall clap their hands." (Isaiah 55:12)

This verse is part of a joyful passage that tells how the things God created will rejoice when His Word goes out into all the world and does good things!

Time to do Activity 25 in the Activity Book!

How Leaves Look Up Close

A microscope is a wonderful tool! It helps us look at things that are too tiny to see with our eyes. Microscopes make small things look bigger.

Leaves are very interesting when we look at them through a microscope. If we look through a microscope, we'll see that leaves are made up of thousands of small things called cells. These cells are attached together. Sometimes they are arranged like bricks or stones in a wall, and sometimes they are arranged like the pieces of a jigsaw puzzle.

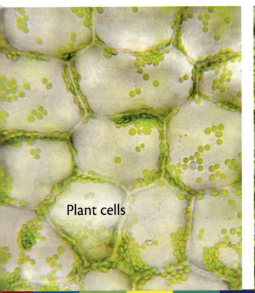
Plant cells

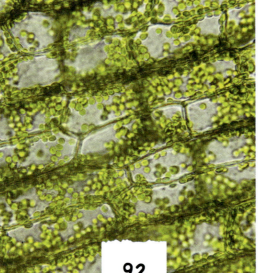

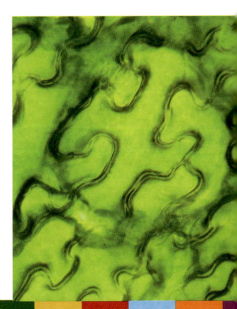

Each leaf cell has liquid and a lot of little working parts inside. Each cell also has a **cell wall** around it to hold it together. Cell walls are made of cellulose to give plants their structure. Underneath the plant's cell wall is a **cell membrane**.

Cell membranes are able to choose what they let into and out of their cell. They might let nutrition and water in. They might let out waste or good things needed somewhere else. Plant cells have both cell walls and cell membranes. People and animal cells have only cell membranes.

Time to do Activity 26 in the Activity Book!

Prayer

Cells are amazing, Lord! You made them so small, too small to see. There are also too many cells to count in a plant or in me. Yet, separately and together, they have such important jobs for all life. Thank You! Amen.

Every living thing is made of cells. That includes you! Cells are often called the building blocks of life.

Today is Green Thumb Day! Time to do Activity 27 in the Activity Book.

CHAPTER 10

Breathing Leaves

W have learned that, during photosynthesis, plants divide water molecules into pieces of oxygen and pieces of hydrogen. They put the oxygen from those molecules into the air for us to breathe. Let's look more closely at leaves to see where this happens.

Tiny Mouths

Leaves have several kinds of cells that do different things. Some of these cells are called **guard cells**. Guard cells have the job of letting moisture and air in and out of little openings on each leaf. There are two guard cells around each opening. Each little opening, with its guard cells, is called a **stoma**, or **stomata** if there are more than one. The guard cells look like two lips of a tiny mouth. They make the stoma open and close at just the right time to do its different jobs.

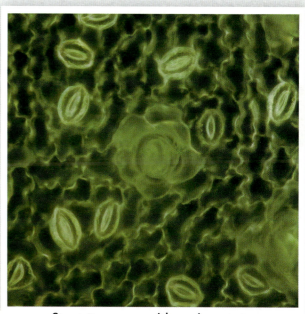
Stomata as seen with a microscope

Stomata are very small. If the period at the end of this sentence was a tiny piece of leaf, there would be about 75 stomata on it. Stomata are mostly located on the shady **underside** of leaves.

Definitions

This book is written in the English language. Long ago, a language called **Latin** was spoken in many places in the world. Scientists have decided to use the Latin language for many science words instead of having different words in different languages. This way scientists around the world can use the same word for the same thing.

Stoma is a Latin word that means "tiny mouth." If we said "tiny mouth" in the Spanish language, we would say "boca diminuta." Each language in the world would have a different way to say "tiny mouth." It's much easier to use just one Latin word, stoma, for this science word!

Herby's List of Stomata Jobs

- Stomata let oxygen gas out into the air during photosynthesis.

- Stomata bring in another important gas called **carbon dioxide** during photosynthesis.

- Stomata let moisture out of the plant.

- Stomata can close to prevent too much moisture from leaving the plant.

Stomata are usually open in the daytime for photosynthesis, but are closed at night. Some desert plants are different. Their stomata open up only in the nighttime coolness so the plant won't lose as much water!

guard cells

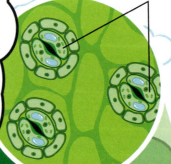

Time to do Activity 28 in the Activity Book!

Corn crops are often given extra water when there hasn't been enough rain.

Tiny Guards

Do you remember that plants put moisture into the air? One corn plant can put two quarts (2 L) of water into the air each day! Water comes up from the roots, travels through the stem, and goes out through the stomata in the leaves. Water doesn't spray or drip out of the stomata as a liquid. It quietly leaves the plant as a gas or vapor, with the tiny pieces of water all spreading out in the air. When water changes from liquid to **vapor**, it's called **evaporation**. Water evaporates out of leaves at the open stomata.

The little guard cells around a stoma can open and close to let out the right amount of water vapor. Stomata must be open during photosynthesis to let gases in and out. But, while they are open, a lot of water vapor goes out too. If more water goes out through the leaves than comes in through the roots, the plant wilts. If the plant wilts, the guard cells will close the stomata. This stops water from escaping. The stomata will remain closed until more water can come up from the roots.

This movement of water from plants to the air is called **transpiration**.

If you forget to water a plant, its cells lose water and get squishy. This makes the plant droop and wilt. It might even die. But if you water the plant again soon enough, your plant's cells will fill up with water and the plant will stand up straight!

Most of the moisture in the air is from transpiration.

Transpiration

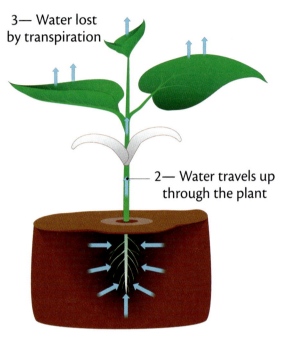

3— Water lost by transpiration

2— Water travels up through the plant

1 — Roots take up water from the soil

Set a guard, O LORD, over my mouth; Keep watch over the door of my lips. (Psalm 141:3)

The little guard cells are careful about what they let out of the mouth-shaped stomata. They remind me of this verse. We need to have a guard for our mouths so wrong words don't come out!

Time to do Activity 29 in the Activity Book!

Prayer

Lord, thank You for making guard cells to be shaped like little mouths. You remind us that our mouths need to be guarded too! We are thankful for the oxygen and moisture You put into the air everywhere through plants. Help our words be good words that praise and thank You everywhere. Amen.

Flower petals don't have stomata because they don't do photosynthesis. But I could still wilt if my plant doesn't have enough water to give me!

Today is Green Thumb Day! Time to do Activity 30 in the Activity Book.

CHAPTER 11
Photosynthesis

W have already learned a little about photosynthesis. We know that photosynthesis is the way plants make food for themselves and for people and animals. We've learned that plants can do this by using sunlight, water, and air.

Little Green Machines

All cells have small parts inside them that have certain jobs to do. The cells on the sunny, upper side of plant leaves have special green machines inside them. These machines are green because of a liquid called **chlorophyll**. Chlorophyll is one of the most amazing things in the world! It can turn sunlight energy into food energy for us. People have never been able to figure out how to do this, but God does it all over the world, all the time, when chlorophyll does photosynthesis.

We know that people and animals breathe the oxygen made by plants. But plants also need to use a little oxygen for certain jobs. During the day, plants can use sunlight to do photosynthesis and make the oxygen they need. But at night, when there is no sunlight, plants must get oxygen from the air. This oxygen can't go through the stomata because the stomata are closed at night. Instead, God made plants with the ability to absorb oxygen right into the cells of the plant!

Flower petals can't do photosynthesis! My pink petals don't have green chlorophyll, so I must get my oxygen from the air around me.

Cells of a moss leaf showing green machines. Without chlorophyll, there would be no life on Earth.

Time to do Activity 31 in the Activity Book!

How Photosynthesis Works

We are going to put together everything we have learned and look at the whole process of photosynthesis. But before we do, let's learn about an amazing symbiosis God made all over the world. Remember, symbiosis is when two living things help each other. Photosynthesis is part of a huge symbiosis. We have already learned about one side of it: it's when all the plants of the world help all the people and animals of the world by making oxygen for them to breathe.

The other side of this amazing symbiosis is that all the people and animals of the world help all the plants of the world by making carbon dioxide for them. We make carbon dioxide in our bodies, but we don't need it. It's a

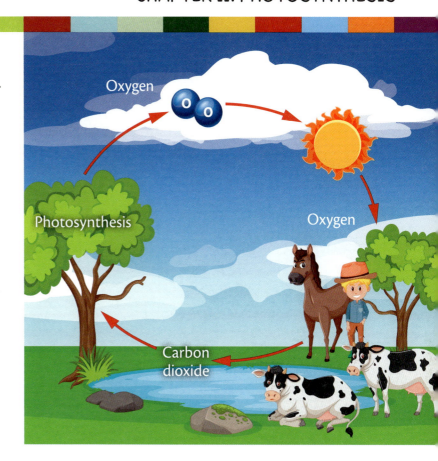

waste product that comes out of our lungs when we breathe. But plants need this carbon dioxide to make food during photosynthesis. God is amazing to make plants able to use the carbon dioxide we don't need, and to make us able to use the oxygen that plants don't need!

As the picture to the right shows, people and animals use oxygen that plants have put into the air. Plants must get carbon dioxide that has gone into the air from people and animals. Plants also need to use a little of the oxygen they make.

The Process of Photosynthesis

1. Water (H2O) travels from the plant's roots to its leaves.

2. Chlorophyll in the green machines of leaf cells absorbs the sun's energy.

3. The chlorophyll uses that energy to split apart water molecules that have come up to the leaves.

4. Oxygen (O) from the split water molecules goes out through the open stomata, two by two, as oxygen gas (O2).

5. Hydrogen (H) from the split water molecules stays in the plant and will be made into food (sugars).

6. Carbon dioxide (CO2) comes into the leaves through the open stomata. Carbon dioxide is made of carbon (C) and oxygen (O) atoms.

7. Carbon atoms and oxygen atoms from the carbon dioxide join with the hydrogen atoms from the split water molecules. They attach to each other and become sugar molecules (food).

Photosynthesis

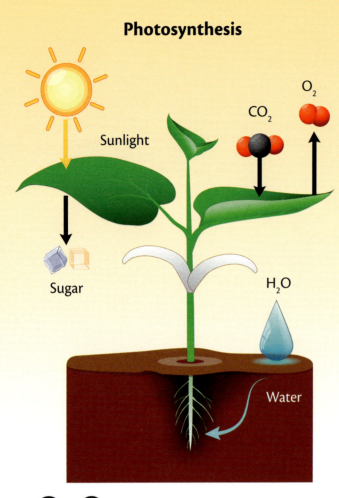

Sunlight

CO_2

O_2

Sugar

H_2O

Water

Where does the work of photosynthesis happen in plants? It happens mostly on the sunny sides of leaves!

The LORD is my shepherd;
I shall not want.
He makes me to lie down in green pastures;
He leads me beside the still waters.
He restores my soul.
(Psalm 23:1-3)

 Time to do Activity 32 in the Activity Book!

Prayer

Thank You, God, that You made a gigantic symbiosis between plants and creatures. We are so happy to know that we will never run out of air. We're also glad we can give plants something they need just by breathing. Thank You for restoring our souls like a shepherd restores his sheep by taking them to rest in green pastures. Amen.

Palm leaves

Isn't it amazing the way plants make food using their little green machines?

Today is Green Thumb Day! Time to do Activity 33 in the Activity Book.

CHAPTER 12

Special Plants in Special Places: Deciduous Forests

About Deciduous Forests

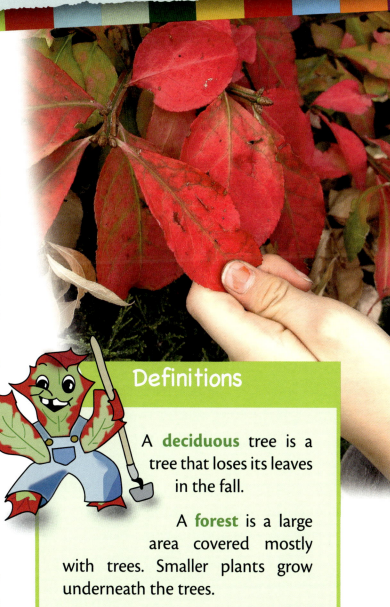

Deciduous forests usually live in a place that has hot summers, cold winters, nutritious soil, and a lot of rain or snow throughout the year. This biome gets more rain and snow than the grassland biome. The thick layers of rich soil in these **forests** are perfect for the deep roots of **deciduous** trees.

The most common plants in deciduous forests biome are trees that lose their leaves in the fall. Their leaves turn lovely fiery colors before they fall off. These deciduous trees then have bare branches all winter long and grow new leaves in the spring.

Why did God make the leaves of deciduous trees fall off? Remember, trees lose a lot of water through their leaves. Deciduous trees don't want to

Definitions

A **deciduous** tree is a tree that loses its leaves in the fall.

A **forest** is a large area covered mostly with trees. Smaller plants grow underneath the trees.

sets earlier and earlier. The tree notices this and makes a special chemical. This chemical makes cells die where the leaf's petiole connects to the stem. This dead layer blocks water to keep it from coming into the leaf.

Without water, and with colder autumn temperatures, the green machines can't make chlorophyll. The chlorophyll that's still in the leaf breaks down, and the green color fades away. Some green leaves have others colors in them, but we can't see these colors in the summer. The green is too dark to let the other colors show. But when the chlorophyll goes away in the fall, the leaves are able to show their other bright, beautiful colors.

When trees lose their leaves, it gives them a rest from photosynthesis too!

This deciduous aspen forest is probably connected underground by the same huge, spreading mass of roots. Each aspen tree is part of the same plant as all the other trees around it. An aspen forest named "Trembling Giant" in Utah, US, is the largest living thing in the world.

lose water all winter long. By dropping their leaves in the fall, trees save water all through the long winter months.

Without any leaves, the trees might look like they're fast asleep. But they are not completely asleep. Their branches spend the winter making buds for next spring's leaves and flowers.

How does a leaf know when to fall? In the **autumn** or fall season, the days get shorter. The sun rises later and later and

Some kinds of oak leaves stay on their trees all winter. Then, in the spring, they are pushed off by the swelling buds of new leaves.

Some leaves have yellow in them, and this yellow begins to show in the fall. Orange is another color that comes out of hiding when the green fades. Leaves that turn brown don't have any pretty colors hiding. As their chlorophyll goes away, the leaves turn brown because they're dying.

Something different happens to make some autumn leaves turn red or purple: the weather! A cold snap will trap sugar in the leaves of certain trees. After that, if there are some days with very strong sunshine, the leaves will make beautiful red and purple chemicals with the trapped sugar.

As autumn moves along, the layer of dead cells between the leaf's petiole and the branch becomes very dry. It breaks easily when a breeze twists the leaf around. The leaf is blown off its twig, twirls through the air, and falls to the ground.

Some years, the leaves look extra colorful because the weather has been just right!

Usually aspen trees only turn yellow in the fall. This picture was taken in a year when the weather caused red color to form in the leaves.

It's fun to run through crispy, colorful, autumn leaves!

 Time to do Activity 34 in the Activity Book!

Special Plants in Deciduous Forests

The plants in deciduous forests grow in five layers.

The **tree layer** has tall trees from 60-100 feet (20-30 m) tall. Oak, maple, hickory, beech, and elm trees are some deciduous trees that get very tall.

Sugar maple tree and leaf in autumn

Beech tree and leaves in spring

Oak tree in spring

Hickory tree and leaves in autumn

110

The **small tree layer** has trees that don't grow tall. Some of these trees are dogwoods, redbuds, and ginkgoes. This layer also has young trees that will get tall years later.

Dogwood blossoms

Ginkgo trees are small Asian trees.

The **shrub layer** has bushes like azaleas and rhododendrons (below).

Redbud blossoms

Azalea bushes make forests colorful under the trees.

Trilliums, ferns, and carpets of phlox grow in the rich soil of deciduous forests.

The **herb layer** has annual wildflowers, ferns, and short plants. Some of these plants do well in the shade. Others prefer the sunshine. The sun-loving plants do their growing and blooming in the spring and early summer before the tree leaves grow large enough to block the sun.

The shortest layer is the **ground layer**. This layer is where mosses and lichens grow.

Time to do Activity 35 in the Activity Book!

This southern beech forest in New Zealand is a perfect place for moss to grow.

Special Animals in Deciduous Forests

From the damp soil to the sunny tops of the trees, forests are full of animal life. Squirrels, chipmunks, mice, deer, raccoons, skunks, and bears make their homes in forests. A deciduous forest near you might have very different animals than a forest far away. Let's look at some special animals in some of these forests!

North American Deciduous Forests

During the warm seasons, broad-winged hawks live in North America's deciduous forests. But in the fall, hundreds of them flock together and migrate to tropical forests in South America.

The groundhog, or woodchuck, is a very large rodent that lives in underground burrows in American deciduous forests.

Eastern screech owls use hollow trees for winter shelters, nesting places, and as cupboards to store extra food when they are too full to finish their dinner.

Luna moth caterpillars eat the leaves of deciduous trees. Adult moths do not eat at all! They only live a week after coming out of their cocoons, just long enough for a mother and father moth to find each other and lay eggs.

Deciduous Forests in Southern Parts of the World

South America, Australia, and New Zealand have forests of southern beech trees. These parts of the world are mostly warm, but the weather does get cold enough in these forests for the beeches to lose their leaves.

God has made the wonderful lyrebird to live in Australia's beech forest. The lyrebird's throat has the most complicated song-making muscles of any bird. Male and female lyrebirds can copy almost any sound. They copy other birds, animals like koalas and dingos, and even noises that come from people, like phone ringtones and the sound of chainsaws!

Lyrebird

Australia has big forests of eucalyptus trees with koalas that like to eat their leaves. But eucalyptus trees are not deciduous. They don't lose their leaves (except to koalas).

European Deciduous Forests

Have you heard of Patrick of Ireland and Alfred the Great? These good Christians lived long ago when Europe was mostly covered with deciduous forests. In one forest, Boniface of Germany chopped down a huge oak tree because people were worshipping it instead of worshiping God.

These European forests were thick and easy to get lost in. The branches and leaves of the trees made a ceiling that didn't let much light through to the forest floor. There were wolves and bears living in the deep darkness.

But the people needed sunny places to grow their food. They would chop down the trees to make farms on the land. Over the years, more and more farms have been made. But there are still many forests in Europe that are the same as they were long ago.

Some European bison still live in specially preserved European forests.

Before hibernating, the hazel dormouse eats enough nuts, seeds, berries, and insects to get two times bigger than its normal size. Then it makes a nest, wraps its tail around itself, and goes to sleep for at least six months!

Asian Deciduous Forests

Siberian tigers, the largest cats in the world, live in some Asian deciduous forests. This tiger is standing on its hind legs to make scratches on a tree trunk. This is the way he tells other tigers to stay away from his neighborhood.

Time to do Activity 36 in the Activity Book!

Prayer

Dear Father, we praise You for the beautiful leaves You gave to deciduous trees: the restful greens, the happy autumn yellows, and the fiery reds. We even like the brown leaves because of the crunchy sound they make as we run through them! Thank You for the beautiful bare branches we can look through to see the winter sky. Your trees give food and homes to the many wonderful animals living in deciduous forests around the world. Amen.

Japanese macaques like the cold forest, but they also enjoy warming up in Japan's hot springs. Young macaques like to make snowballs to play with.

UNIT 4
Stems: The Sturdy Pipes of a Plant

 ## Memory Verse

Let the field be joyful, and all that is in it.
Then all the trees of the woods will rejoice before the LORD.
For He is coming.
(Psalm 96:12-13)

This memory verse and hymn are both joyful! They say that God's creation, like fields and trees in the woods, rejoice and sing because the Lord is coming.

 ## Hymn to Sing

Joy to the World

Joy to the world! the Lord is come:
Let earth receive her King;
Let every heart prepare Him room,
And heav'n and nature sing,
And heav'n and nature sing,
And heav'n, and heav'n and nature sing.

Joy to the earth! the Savior reigns:
Let men their songs employ;
While fields and floods, rocks, hills, and plains
Repeat the sounding joy,
Repeat the sounding joy,
Repeat, repeat the sounding joy.

No more let sins and sorrows grow,
Nor thorns infest the ground:
He comes to make His blessings flow
Far as the curse is found,
Far as the curse is found,
Far as, far as the curse is found.

You can listen to this carol by searching for "Joy to the World for kids" on the internet.

CHAPTER 13
Sturdy Stems

Kinds of Stems

W usually think of plants in big groups: trees, bushes, herbs, climbers (vines) and creepers. These groups are different from each other because their stems are different.

Trees have one woody stem called a **trunk**. Trees are sturdy and can grow tall and large because their stems are wood. They are perennial plants and stay alive aboveground all year long for many years. Trees are some of the oldest and largest living things!

A vine is a curly stem that can grab onto things and lift itself up as it grows. Grapes grow on vines that must have their branches cut (or pruned) at the right time in order to make lots of grapes. Jesus says He is a vine. Wouldn't you like to be a fruitful branch for Jesus?

"I am the true vine, and My Father is the vinedresser. Every branch in Me that does not bear fruit He takes away; and every branch that bears fruit He prunes, that it may bear more fruit." (John 15:1-2)

Most trees have woody branches. Palm trees have leafy branches that form only at the top of the tree.

In science, **herbs** are all the plants that have soft, green stems instead of stiff, woody ones. The stems of herbs are able to stand up because their cells are full of water. If a plant gets too dry, the cells in its stem lose water, and it wilts. Some herbs are annuals that completely die each year. Other herbs are perennials whose roots stay alive even though their aboveground parts die each year. Herbs can be either narrow-leaved **grasses** or wide-leaved **forbs**. Herbs are smaller than shrubs.

Grasses have leaves that look like lines.

A **shrub** (or bush) has several thin **stems** coming out of the ground instead of just one thick trunk. Shrub stems are woody and do not die aboveground each year.

Forbs have wide leaves of different shapes.

Grape tendrils help their vines hang on as they climb.

Climbers (vines) are plants whose stems are too weak to hold the plant up. Instead, they grow upward by attaching themselves to something sturdy as they climb. Climbers often have **tendrils** along their stems that they use to grab onto their sturdy support. Tendrils are special stems.

Pumpkin creeper sprouting roots to get more water

Creepers also have weak stems, but they don't try to climb upward. They creep sideways along the ground, often sprouting new roots as a way to get more water or make a new plant.

Strawberry creeper sprouting new plants

Strange Stems

Bulbs and tubers are both underground stems that grow fat with the food they are storing. **Bulbs** are usually rounded with a pointed top and roots coming out the bottom. **Tubers** are not round. They have little buds called **eyes** that new plants can grow from.

Eve a beautiful garden to take care of. The world was perfect then. But Adam and Eve sinned by disobeying God. God punished them by putting some curses into the world. The world would no longer be perfect.

One of the curses was death. Adam and Eve and all the people who came after them would one day have to die. The other curses made life harder. In this verse, God tells Adam how his life would be harder in this changed world:

**"Cursed is the ground for your sake;
In toil you shall eat of it
All the days of your life.
Both thorns and thistles it shall bring forth for you."
(Genesis 3:17-18)**

Onion bulbs and potato tubers are two kinds of underground stems.

Let's see what God says about **thorns** and thistles. When God created the world, and filled it with good things, the last things He made were people. The first two people were Adam and Eve. God gave Adam and

Thorns are special stems that protect a plant. Ouch!

This verse tells us why we have thorns and thistles. It's sad that Adam and Eve didn't obey God. But it's also sad that we don't always obey God either. We must all live in this world that isn't perfect, and we must all die someday.

But there is wonderful, joyful news! Our hymn "Joy to the World" sings of the good news we read in the Bible. This good news is that Jesus came to forgive our sins and give us eternal life. Life might be hard with thorns and thistles, but Jesus came "to make His blessings flow far as the curse is found."

 Time to do Activity 37 in the Activity Book!

Wonderful Wood

Wood is an amazing invention of God! Its sturdy strength is very useful to us. Trees and bushes are different than other plants because they have stems made of wood.

Do you remember that plants and trees have special growing places at their tips? This is the place where they make more cells so they can get taller. And do you also remember that God gave grasses special growing places near the ground so they can get taller even if an animal eats the top off the plant?

God has given trees and bushes more special growing places. They have special growing cells under all their bark that make them get fatter! Trees and bushes get *taller* only at the tips of their trunks

and branches. But they can get *fatter* under the bark all along their trunks and branches.

This special growing area under the bark is called the **cambium**. The reason cambium makes tree and bush stems fatter is because it only grows sideways.

God has given the cells of the cambium the special gift of being able to change and become different things. At the tips of all plants, the cells in the growing places can become either leaves, flowers, or more special growing cells. In the cambium, the special cells can become bark toward the outside of the tree, wood toward the inside, or more special g r o w i n g cells in the cambium.

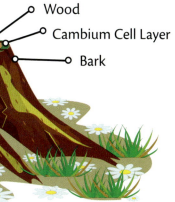
Wood
Cambium Cell Layer
Bark

Bark often cracks because the tree inside is growing fatter.

The cambium makes new bark and new wood only during the growing season, not in the winter. In the rainy months of the growing season, the new wood is light colored. During the growing season's dry weather, the new wood is darker. This makes light and dark **tree rings** in the wood. We can see tree rings when a tree is cut down and we look at the round stump.

The bark of aspen trees often looks greenish. That's because it can do photosynthesis, even in winter! These aspen trees show marks on their bark where elk scraped them with their teeth. The elk were glad to find green food when snow had buried all the herbs.

Sometimes people try to tell how old a tree is by counting its rings. But this isn't always exact because there may be several wet and dry times in one growing season. This would show up inside the tree as several light and dark rings in the same year.

The righteous shall flourish like a palm tree,
He shall grow like a cedar in Lebanon.
(Psalm 92:12)

Cedar trees can get very large. Their trunks have tree rings.

Palm trees grow in areas of the world that don't have much change of weather in the seasons. Their trunks don't have tree rings.

Woody stems are amazing because they have a lot of **lignin**. Lignin is a wonderful chemical that goes between the cellulose fibers of cell walls. It makes them stiff and strong. Lignin also goes between the cells to make them stick tightly together. Trees would topple over if it weren't for lignin. Lignin is what makes wood strong.

Different kinds of wood have different smells when they burn. That's because each kind of tree has a slightly different kind of lignin. People make smoked meats with different tastes by burning different kinds of wood under the meat.

Prayer

Dear Father, You have made every stem just right for holding up its plant. Tree and bush stems are very study. Herb stems are sturdy enough to stand if they have enough water. And climbers and creepers have the stems they need so they can grow in the best places! Thank You for making so many different plants that are able to live in so many different places. Amen.

Time to do Activity 38 in the Activity Book!

In parts of the world, bamboo is plentiful. Bamboo stems provide bowls for this feast.

Sequoia trees have some of the tallest stems (trunks) in the world! The General Sherman Tree is the world's largest tree. It stands 275 feet (83 m) tall. Its trunk is over 36 feet (11 m) wide at the bottom.

CHAPTER 14
The Pipes in Stems

Have you ever thought about all the pipes in your house? Do you know that some pipes bring clean water into the house and carry it to where it's needed? Plants also have tiny pipes to bring clean water in from the soil and carry it all around the plant.

Your home has other pipes that take out dirty water. Plants don't do this. But they do have another kind of pipe that is not like pipes in your house. This other kind of pipe picks up good things the plant makes and carries them around to other parts of the plant.

Going Up!

We've learned that plant roots

Let's learn about these interesting plant pipes and the work they do!

absorb water from the soil. A little of that water travels up to the leaves and is used in photosynthesis. With transpiration, the rest of that water travels up and out of the plant through the stomata while they are open during photosynthesis. Some trees are

very tall. How can water travel all the way to the top of a tall tree even though gravity is pulling it down?

The trees of the LORD are full of sap, The cedars of Lebanon which He planted. (Psalm 104:16)

The Bible says the trees of the Lord are full of **sap**. Sap is the water (and the food and minerals it carries) inside plants.

God has made a way for liquid to travel upward in plants through a lot of tiny pipes called **xylem** (zye-

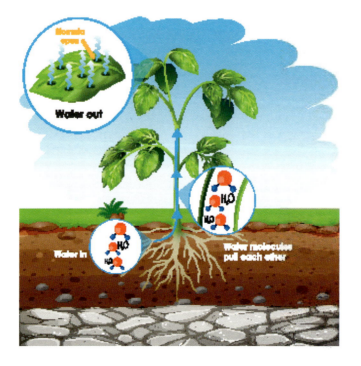

lum). Each xylem pipe is made up of a long chain of cells. These cells die, become stiff, and open up to make the pipe. Water travels through these pipes, but the dead cells can't do anything to help the water move up. Instead, water travels upward because of the way God made water!

Water molecules like to stick to things. If water molecules are in a skinny pipe, they stick to its walls and climb up a little way. Xylem pipes are skinny enough for water to do this. Scientists think that this helps water climb.

But water can only climb a little way up skinny pipes. It wouldn't be able to climb all the way to the top of a tall tree. It needs help, and it gets this from transpiration. Scientists think that transpiration is the reason water can climb so high in the xylem.

Besides sticking to things, water molecules like to stick to each other. They stick to each other because the atoms of one water molecule are attracted to the atoms of another water

Plants can bring water up through their xylem straws.

of the open stomata, it evaporates and flies away! As it does, it pulls up another water molecule to take its place. All the other water molecules are pulled up behind it because they like to stick together. This is like when you drink through a straw. As you suck some liquid into your mouth, more liquid must travel up the straw to take its place.

molecule. During transpiration, the water molecules in the xylem hold onto each other from the roots to the leaves. As one of the water molecules reaches one

Besides water, the xylem also carries up minerals from the soil!

Time to do Activity 40 in the Activity Book!

Going Down and All Around

Plants have different pipes to carry good things down, up, and around in the plant. They are called the **phloem** (flow-um). Cells all over the plant are doing important jobs that take energy! They need the food energy that's been made in the leaves. The phloem delivers that food and other needed things all around the plant.

Phloem pipes are made up of chains

of *living* cells. The cells need to be alive because their job is to keep the liquid goodness from rushing straight

Definitions

Xylem is all the tiny pipes that carry water and minerals up from the soil.

Phloem is all the tiny pipes that deliver good things made by the plant to the parts of the plant that need them.

133

down to the roots. Phloem cells have walls separating them, but these walls have little openings (**perforations**) to gradually let liquid through. Phloem pipes have other cells surrounding them that help deliver nutrition from the phloem to the right parts of the plant.

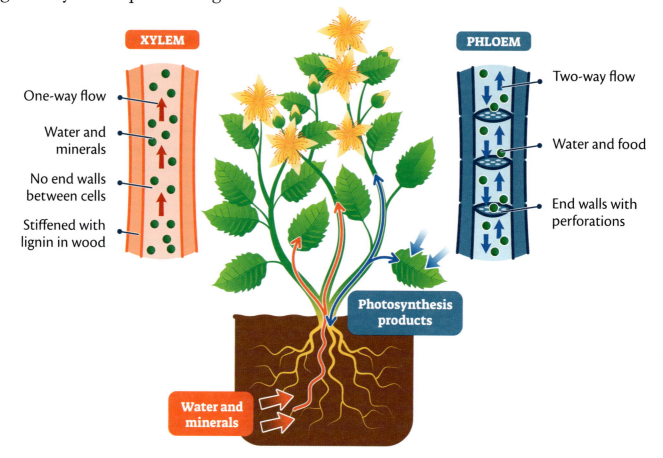

Prayer

Oh God, You can do such big things in such tiny ways. Xylem and phloem pipes are so small, but without them, there would be almost no plants on Earth. These pipes are too small to see without a microscope, but they have such important jobs. We praise You for making them! Amen.

Time to do Activity 41 in the Activity Book!

Unusual stem

Vines are climbing stems that sometimes lose their hold and become fun ropes!

CHAPTER 15
Stems in Motion

We've learned that phloem carries food from the leaves and delivers it to the rest of the plant. But phloem also carries chemicals that give messages to the plant.

Do you remember that plants know when to start making flowers because a special chemical takes the message from the leaves to the growing tip of the plant? Special chemicals like these are called **hormones**. Hormones are chemical messengers made in one part of a plant, person, or animal. When the time is right, a hormone takes its message to another part of the plant or creature and tells it to do a certain job.

Auxin Tells Stems to Move!

Let's learn about some plant hormones!

Auxin (ox-in) is a plant hormone that tells the cells in stems to grow longer. When a stem's cells get longer, the stem gets longer. The plant grows!

Besides growing, God uses auxin to help plants move in other ways:

1. **Plants can lean toward light.** If a plant is indoors next to a window, light will only shine on one side of the plant. On the other side, its leaves will be in the shade where they can't do photosynthesis very well. But after a few days the plant's stem may have bent so the leaves on the other side of the plant can reach the light. This happens because auxin is made at the top of the plant and is sent down to the shady side of the stem. The auxin makes

the shady side of the stem grow faster than the sunny side. Because of this, the stem leans toward the light! And, because of auxin, the leaf petioles also bend so their leaves face the light.

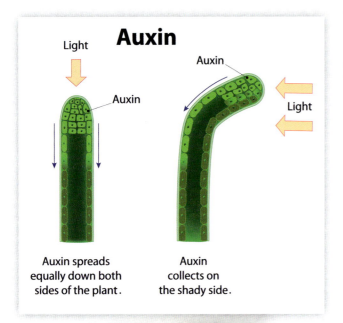

Auxin

Light

Auxin

Auxin

Light

Auxin spreads equally down both sides of the plant.

Auxin collects on the shady side.

Mallow leaves turn to face the sun all day long. After sunset, the leaves turn to the east so they will be facing the sun when it comes up again the next morning! Isn't God amazing?

Auxin also helps outdoor plants lean toward the light. Some leaves and flowers turn to face the sun all through the day.

"Repent, then, and turn to God, so that your sins may be wiped out, that times of refreshing may come from the Lord." (Acts 3:19 NIV)

When stems, leaves, and flowers turn toward the light, it reminds me that we should turn away from our sins and turn to God. Because of Jesus' death, our sins will be wiped out! We will be refreshed like a wilted plant that finally gets water.

Snow buttercups bloom when winter is barely over! Their flowers are about three degrees warmer than the air around them because they turn to face the sun all day long. In the chilly spring, pollinators are attracted by the warm yellow flower cups.

2. **Plants know which way is up.** When a new stem pokes out of a sprouting seed, it might be pointing upward, downward, or sideways. But soon gravity causes auxin in the little stem to settle in the lowest places. Where auxin settles, the cells get bigger. These bigger cells make the stem lean away from the pull of gravity and move up through the dirt as it grows.

touches something, the auxin moves away from that side of the stem. That means there is more auxin on the side not being touched and it causes those cells to get longer. When those cells get longer, the stem leans towards the thing it's touching and eventually wraps itself around it.

Tendrils are special stems on vines. These dried pea tendrils have curled around a fence to support their high-climbing pea vine.

3. **Plants can grab things.** God has given climbing plants an interesting way to wrap themselves or their tendrils around something sturdy to climb as they grow. Auxin helps with this too! When a vine or its tendril

Time to do Activity 43 in the Activity Book!

139

Other Hormones Take Messages through the Stems

There are several other plant hormones that travel through the phloem to give messages:

Okay, cells, it's time to use your germ-fighting weapons!

1. Cytokinin helps plants grow by telling cells to divide and make more of themselves. It also tells the leaves when to make more green machines.

2. Gibberellin tells a plant when to start growing in the spring and when to make flowers. It also can tell seeds when to sprout.

3. Abscisic acid tells the plant when to stop growing in the fall. It can also tell the stomata to close if the plant's roots get too dry. This hormone also makes leaves fall and keeps seeds from sprouting at the wrong time.

4. Ethylene is a hormone that tells fruit when to ripen. Often bananas are picked unripe and green. Then, before they are sold, they are sprayed with ethylene to make them ripen.

5. There are other hormones that are danger signals. If a disease is infecting one part of the plant, a certain hormone sends a signal to the whole plant, telling its cells to change into germ-fighters.

6. If an insect or animal takes a bite of a plant, a special hormone will give a danger signal telling the whole plant to make itself taste bad.

7. God also made an amazing plant hormone that can go out into the air and call female wasps to come help. When a plant is being eaten by caterpillars, the plant puts out chemicals that attract certain wasps. These wasps will lay their eggs in the caterpillars' bodies! The baby wasps that hatch will eat the caterpillar from the inside and kill it. Scientists did experiments and learned that the chemicals did not come from the caterpillars. Also, the chemicals weren't made when the scientists crushed the leaves. The chemicals were only made by the leaves when *caterpillars* were eating them!

After feeding inside this tomato hornworm on the right, wasp larvae crawled to the outside and spun cocoons around themselves. Adult wasps will soon emerge from the cocoons, and the hornworm will die because of the damage done earlier by the wasp larvae.

Prayer

Lord, thank You for giving plants special hormones that help them grow and change at the right time. We especially praise You for the way plants can defend themselves from disease and from being destroyed by insects and animals. You made these hormones work so perfectly! Amen.

Time to do Activity 44 in the Activity Book!

Hormones do a lot of work to help plants. And the phloem keeps busy delivering them!

Today is Green Thumb Day! Time to do Activity 45 in the Activity Book.

CHAPTER 16

Special Plants in Special Places: Coniferous Forest

About Coniferous Forests

The most common plants in the coniferous forest biome are **conifer** trees. Conifer trees, like pine, spruce, and fir, don't have flowers or fruits. Instead, they make cones for their seeds to grow in.

If you were to take a walk in a coniferous forest, you would always remember its beauty: the trees, so tall and strong, with sunlight filtering through the green or with snow collecting on the branches; the forest floor with its comfortable cushion of old needles for you to quietly walk on; and the clean smell of the needles and bark when you rub them with your fingers. You might remember the silence. Or if the wind is blowing, you might hear soft sighing in the branches high above. And you would be on the ground below, sheltered within the thick woods, amazed at what God has made.

Not many people live in the largest coniferous forests of the world. These forests are in the north: in Canada, Alaska, Scandinavia, and Siberia.

Log cabin in one of Canada's coniferous forests

These northern forests are usually called the **taiga** (tie-guh). Taiga land is often flat. Rainwater and melted snow collect in lakes, ponds, and in wet, spongy areas called **bogs**.

The coniferous forest biome has long, cold winters and short, cool summers. Conifer trees do well in cold, snowy climates. Thankfully, many people can take walks in conifer forests that are farther south of the taiga. These forests are usually found in mountain areas where the climate is cold. But some conifers can also grow in very warm areas.

Conifers grow well in poor, sandy soil where there isn't enough nutrition for deciduous trees. Deciduous trees need good soil so they have enough nutrition to make a new batch of leaves each year. But coniferous trees are usually **evergreen**. They can grow in poor soil because they don't lose their leaves all at once. Each coniferous **needle** may live for several years before falling off!

The taiga gets about the same amount of moisture as the grasslands. But the cold temperature of this coniferous forest keeps water from evaporating as quickly as it does in grassland. This way, the trees get as much water as they need.

Water in the soil of a coniferous

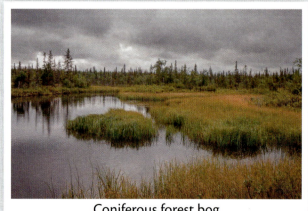

Coniferous forest bog

forest can stay frozen for much of the year. This might seem like a problem because plants can't take in frozen water. Thankfully, conifer trees can take in water when it's liquid. Then, they are able to hold onto the water because they have needles instead of leaves. God made the needles of conifers narrow and covered with wax so they don't lose water easily. Even though the trees' needles stay green all year long, they don't do photosynthesis in the winter. But they can start photosynthesis early in the spring without waiting to grow new leaves like a deciduous tree would need to do.

The soil in coniferous forests is thin. Why is that? Soil is partly made of rotten plants. It takes conifer needles a long time to rot and become soil in such cold temperatures. Under the thin soil, there is usually rock or frozen soil that never melts. Land like this wouldn't work for the deep roots of most deciduous trees. But God made conifers with shallow, spreading roots that are just right for shallow soils.

God made coniferous branches strong enough to hold heavy snow.

This fallen conifer tree shows its shallow, spreading roots.

Time to do Activity 46 in the Activity Book!

Special Plants in Coniferous Forests

Let's learn about some coniferous forest plants! Most of the plants in coniferous forests are trees that make seeds in cones instead of within a fruit.

Coniferous trees don't have flowers, but they do have two types of cones used for seedmaking. They have pollen-making cones (**pollen cones**) and pollen-taking cones (**seed cones**). Conifers are pollinated by the

Definitions

An **evergreen** tree has leaves that stay green all year long. Evergreen trees do not lose their leaves all at once in the fall. They lose their leaves (or needles) a few at a time, but the tree is never bare.

A **conifer** is a tree that makes its seeds in a cone instead of in a fruit.

pollen cone
young seed cone
mature seed cone
seed cone

wind. The pollen cones make plenty of pollen! That way there will be enough pollen blowing around to land on the seed cones where seeds will be made.

Most evergreen trees

It may take three years for seed cones and their seeds to grow up. Seed cones point up when they are young so they can catch falling pollen. When their seeds are grown, cones open to let the seeds sail away on their little wings.

These larch trees have needle-like leaves that turn yellow and drop in the fall, but they make cones like other conifers.

are conifers, and most conifers are evergreens. We can use either word to describe pine, spruce, and fir trees. But there are a few evergreens, like holly, that are not conifers. Holly bushes and trees have leaves that stay green all year, but they don't make cones. There are also a few conifers, like larch, that are not evergreen. Although larch trees make cones, their needles turn yellow and drop in the fall.

"I am like a green cypress tree; Your fruit is found in Me." (Hosea 14:8)

Giant sequoia trees are a type of cedar. They help us think of how big God is. He is the one who gives us what we need!

Fir trees have soft, flat needles.

Spruce needles are stiff and sharp. You can roll one between your fingers because it isn't flat.

Pine needles are grouped in sets of more than one. The set is held together with a sheath near the branch it's attached to.

The leaves of cedars are scales that overlap to cover the twig.

The wood of coniferous trees is softer than the wood of most deciduous trees. Conifers provide all the world's softwood for building. Their long, straight trunks are perfect for making lumber. Paper, cardboard, and pencils are also made from conifer trees.

When the needles of conifer trees finally rot into the soil, they make the soil unfriendly to other plants. Not many kinds of plants can grow under conifer trees.

Mosses *are* able to grow under conifer trees. Cushions of moss are made of thousands of tiny plants. These plants are so tiny that they don't even need xylem and phloem to carry good things around inside them. Instead, water and nutrition are absorbed directly into each part of the moss plant from its surrounding neighborhood. The tiny leaves of moss are only one cell thick!

On this tree farm, you can see one area where all the trees have been cut, another area with replanted young trees, and one tall "seed tree." Long ago, before tree farmers owned the land and replanted trees, loggers would leave a few seed trees. Seed trees would make cones with seeds to replenish the forest with trees. Tree farmers don't need the old seed trees any more, but they often leave them standing as a part of history.

148

Home builders use lumber from conifer trees.

Mossy forest

Isn't it fun that God made a flower and an animal that look alike?

Elephant head flowers bloom in the moist meadows of some coniferous forests.

Sometimes, when conifer trees have been cut down or burned up by a forest fire, sun-loving forbs and grasses grow in the bare areas. After a time, deciduous trees, like aspen or birch, sprout and become a forest there. Eventually, these deciduous trees use up the nutrition in the thin soil. Then they die, and conifer trees are able to take their place.

Fireweed is one of the first plants to spring up after a forest fire.

Time to do Activity 47 in the Activity Book!

Spotted owls like to live in shady forests with big, old trees.

Pine marten

Red squirrel

Special Animals in Coniferous Forests

**For every beast of the forest is Mine.
(Psalm 50:10)**

Pine martens, squirrels, spotted owls, pine beetles, and crossbill birds are some of God's animals that live mostly in coniferous forests.

Pine martens are a kind of weasel. They mostly eat small mammals but sometimes eat plants. When they eat fruit, the seeds pass through their bodies and can still grow into plants. In fact, wild blueberry seeds that have passed through a marten will sprout better than seeds that fall off the plant.

Red squirrels each have a favorite tree where they bring cones and break them apart to eat the seeds. After a while, this makes a big pile of broken cones called a **midden**.

Gray jay

Stellar's jay

Steller's jays and gray jays both live in coniferous forests. Gray jays are also called "camp robbers" because they are bold enough to take food off your picnic table while you are sitting there!

God gave crossbills special beaks. The upper and lower beaks cross over each other! This looks funny, but it has an important purpose: It allows the crossbill to pry open cones to get to the delicious seeds inside.

Pine sawyer beetles can spread tree-killing diseases as they feed on conifers.

A baby porcupine is called a porcupette.

Deer, moose, bears, porcupines, foxes, and mice are other animals that can be found in coniferous forests.

Prayer

Dear Lord, thank You for the beauty of coniferous forests. Thank You for making such big and special trees to grow, even where the soil is poor. We know You can also do wonderful things with us even though we have poor things to offer. Amen.

Time to do Activity 48 in the Activity Book!

Moose are usually seen in or near water within conniferous forests.

These children are climbing on the roots of a fallen tree. Trees don't fall very often, but as this one fell, its roots were pulled out of the ground and tipped sideways. Notice that new trees have sprouted from the old tree's roots! This only happens with certain kinds of trees.

UNIT 5
Roots: Hidden, Humble, and Important

Now we're going to learn about good, sturdy roots that hold plants in the good, sturdy ground. This will help us understand our memory verse and what it means to be rooted in Jesus!

Hymn to Sing

He Formed the Deeps Unknown

He formed the deeps unknown,
He gave the seas their bound;
The wat'ry worlds are all His own,
And all the solid ground.

Come, worship at His throne;
Come, bow before the Lord:
We are His works, and not our own;
He formed us by His word.

This hymn may be sung to the tune of "Row, Row, Row Your Boat," which can be found on the internet.

Memory Verse

As you therefore have received Christ Jesus the Lord, so walk in Him, rooted and built up in Him and established in the faith, as you have been taught.
(Colossians 2:6-7)

CHAPTER 17

What Do Roots Look Like, Hiding Down There?

Roots are the part of a plant that usually lie hidden below the surface of the soil. In most plants, the root growth underground is as large as the plant's growth aboveground. Let's see what a plant looks like down where it's dirty and dark!

Kinds of Roots

Some kinds of plants have **fibrous** roots. Other kinds have **taproots**.

Fibrous roots are thin, branched, and grow right out of the stem of a plant. These roots usually grow close to the surface of the ground and spread sideways instead of growing deeply downward. The fibrous roots of a plant are all the same size.

A taproot is a single root that grows deep into the soil. It forms smaller branches on itself as it lengthens. These branches spread widely and deeply to attach the plant firmly in the earth. God gave trees taproots to prevent them from being blown over too easily.

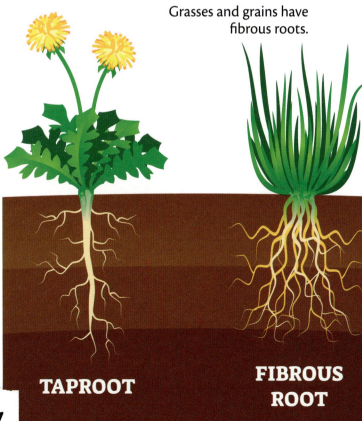

Grasses and grains have fibrous roots.

TAPROOT

FIBROUS ROOT

157

Carrots and beets are special kinds of taproots that store food.

Carrot and beet plants are not annuals, and they aren't perennials! They are **biennials**! That means it takes two years for them to sprout, make seeds, and die.

This beet plant grew one summer and was left in the ground through the winter. Its leaves died, but the root stayed alive. In the spring, the stored food in the beet root was used to grow a new plant. This second plant is now flowering and will make seeds and die.

Time to do Activity 49 in the Activity Book!

Roots Up Close

The tips of roots are where most of the work happens. Let's learn the names of some root tip parts to help us understand the important jobs that God gave to roots!

The **root cap** is a tough covering for the tip of the root. It protects the growing root from wearing down as it plows through soil.

The **area of dividing cells** is a special growing place. These cells divide and become more cells. They can become more root cells or more root cap cells depending on what the plant needs.

Above the dividing cells is the area

Chili pepper plant root

where the root grows by making its cells longer. This is also the very important area where the skin of the root absorbs water and nutrition from the soil.

Root hairs are a part of these root skin cells. Since root hairs can absorb good things, they help the root take in much more water and nutrition than if there were no root hairs.

Xylem and phloem are also made in the roots as they grow. These tiny tubes are connected to the same xylem and phloem of the plant's stems and leaves.

Root Structure

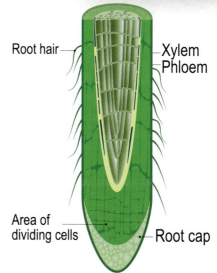

Root hair

Xylem
Phloem

Area of
dividing cells

Root cap

Time to do Activity 50 in the Activity Book!

Prayer

Lord, thank You for fibrous roots that tangle up in the soil to make mats of grass near the surface. Thank You for taproots that grow deeply and securely in the earth and reach to find good things. Thank You for giving roots the right kinds of cells for the work they do. Amen.

Orchids do not have underground roots. Their roots get good things from the rain, the air, and the tree's surface.

CHAPTER 18

What Do Roots Do So Secretly Underground?

Roots Grow in Search of Good Things

A root's main job is to provide the rest of the plant with enough water and nutrition to grow. The roots do all their absorbing near their tips where the root hairs are found. Often, the soil around the root tips becomes poor because the roots have taken out all the good things in the soil. But the root tip solves this problem by growing longer and reaching a new spot where the soil is nutritious.

Have you ever been outside during a rainstorm and stood under a tree for shelter? Maybe a little rain fell through the tree and onto you. But most of the rain falling on the tree would have bounced from leaf to leaf until it fell off the tree onto the ground near the outside edges of the tree's branches. This area of the ground where rain falls under the edge of the tree is called the **dripline**.

God is so amazing because He puts a tree's root tips right under its dripline. This is the way trees can absorb the most water!

> This verse talks about a tree's roots growing toward the water it needs. God wants us to trust and hope in Him so that we will have what we need!

"Blessed is the man
who trusts in the LORD,
And whose hope is the LORD.
For he shall be like a tree planted
by the waters,
Which spreads out its roots by
the river,
And will not fear when

When things are just right, roots can grow as fast as half an inch (1 cm) a day!

heat comes."
(Jeremiah 17:7-8)

Roots are interesting because they know they should grow downward! Gravity is the pull that holds things down on the earth. In the middle of each root cap are some special cells that can sense gravity. These cells use gravity to figure out which way is down, and they send the root growing in that direction.

Do you remember that the hormone auxin tells stems to grow upward? The same hormone tells roots to grow downward! God has put opposite instructions in roots and stems, telling them which way to grow when they sense gravity.

Time to do Activity 52 in the Activity Book!

Roots Pick and Choose

As roots grow longer, the xylem tubes are made longer too. This new xylem in the roots becomes part of the xylem throughout the plant. Water enters the new xylem from the soil and is carried all the way up through the plant. But as water is absorbed into the roots, it can't go directly into the xylem. First, it is stopped by a waterproof

layer of cells. These special cells have the job of filtering the water before letting it into the xylem. The cells can choose to filter out dangerous things, like too much salt, and leave those behind. The special cells can also choose the right amount of good minerals to let through. Then the nutritious water can enter the xylem and travel to the rest of the plant.

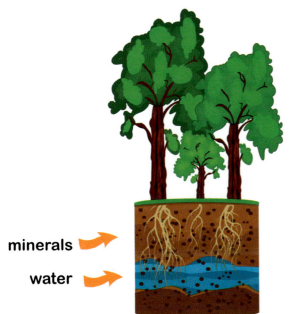

minerals

water

Roots can also communicate with the stomata in the leaves! This happens when the roots sense that the soil doesn't have enough water for transpiration. The roots know the plant could die if all its water goes up through the xylem and out through the stomata. So the roots use the hormone abscisic acid to send a message to the stomata, telling them to close. God takes care of the plants!

Time to do Activity 53 in the Activity Book!

Prayer

Roots are so amazing, God! Even though they don't look very pretty, You have made them able to do such important, complicated things. We praise You! Amen.

CHAPTER 19
Plants Need Soil!

Then God said, "Let the waters under the heavens be gathered together into one place, and let the dry land appear"; and it was so. And God called the dry land Earth. . . . And God saw that it was good. (Genesis 1:9-10)

God made dry land, including soil, on the third day of creation. He made plants on the same day! He made them together because plants need soil. Soil provides a home for a plant. It provides a place for its roots to be anchored and protected from heat and cold. And, although plants make their own food (sugars), they need to stay healthy by using the nutrition that soil can give.

What Is Soil?

Soil is the upper layer of Earth where plants grow. Good soil is a mixture of clay, tiny rock pieces, and rotten plant and animal parts.

Sometimes soil is carried away by water or blown away by wind. This is called erosion. God knew the soil He made would erode, so He also made a way for soil to be renewed.

Soil is renewed when wind and water break off tiny pieces of rock. Sometimes plant roots grow into cracks in rocks and, as the roots get fatter, pieces of the rock break off. Water can also run into cracks in rocks. When this happens in a cold climate, that water will freeze. Water expands as it freezes and often causes rocks to break. All of these pieces of rock fall to the ground and help make more soil. The largest pieces of rock making up soil are called gravel. The

tiniest pieces are **clay**. In between are **sand** and **silt**.

Soil has more things in it than these four sizes of rock. As plants and animals live on soil, they leave behind dead cells and other pieces of themselves. Over time, these once-living things rot and mix with the rock pieces. Small, live creatures come to live in the soil. If you scrape off the top inch of soil from a square yard (meter) of ground, there could be thousands of kinds of visible and microscopic creatures in that soil. This mixture of non-living rock pieces, once-living rotten things, and now-living creatures makes good soil for plants.

If you were to dig a deep hole, you would see separate layers of soil. The **humus** (hew-muss) layer on top is made of completely broken down plant and animal parts. **Bedrock** on the bottom is

Nutrients

Organic Matter

Fungi   Bacteria

Protozoa   Springtail   Nematode   Mite

Centipede   Spider

Earthworm   Ground beetle   Millipede   Ant

the rock of the earth's crust. As you go from humus to bedrock, each layer has more rock material and less humus.

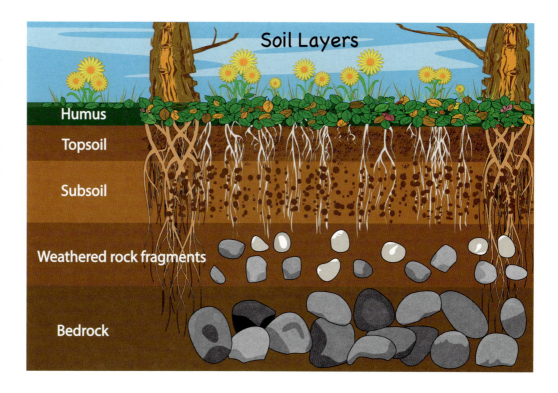

Soil Layers

Humus

Topsoil

Subsoil

Weathered rock fragments

Bedrock

Time to do Activity 55 in the Activity Book!

Why Do Plants Need Soil?

Do you remember that God put sunlight, water, and air all over the earth for His plants to make food? Soil is another amazing thing that God put all over the earth for plants.

When you eat food, you get energy from it. But in the food, you also get vitamins and minerals to give you health. A plant makes its own food for energy, but it needs to get extra nutrition from the soil.

As earthworms eat their way through the soil, they help break down plant material and improve the soil.

171

Herby's List of Good Things Plants Get from Soil

- The **minerals** plants need come from the small rock pieces in soil. Water carries minerals from the soil into the plant's roots.

- Plants need a chemical called **nitrogen**. There's a lot of nitrogen gas in the air, but it's not in a form plants can use. Certain microscopic bacteria in soil can enter a plant's roots and change the air's nitrogen to a form that the plant can use. In return, the bacteria are given sugars from the plant's photosynthesis. There is symbiosis between the plant and the bacteria!

- A special hair-like **fungus** exists in soil and grows in and around plant roots. It absorbs water and nutrition from the soil and gives it to the roots. In return, the plant gives the fungus sugars it has made. There is symbiosis between the plant and the fungi! Most plants would not be able to get enough nutrition from the soil on their own and would die without these helpful microscopic fungi.

- Plant roots need **air**. The empty spaces in soil hold air for the roots' oxygen needs.

- **Water** moves and spreads slowly through soil, giving plant roots time to absorb it.

- The soil is also home to many creatures. As they eat, dig, and live, the creatures break down dead plants, giving **usable nutrition** to the roots of living plants. The soil gives the creatures protection and a place to find food.

Time to do Activity 56 in the Activity Book!

Definitions

A **mineral** is a solid chemical that comes from things that were never alive. Rocks and metal are made of minerals.

A **fungus** is a kind of living thing that gets its food from rotting plants and animals. If we are talking about more than one fungus, we say fungi (fung-guy). Fungi act a little like plants, but they don't have green machines and don't do photosynthesis. Mushrooms, yeast, and mold are kinds of fungi.

Prayer

Heavenly Father, You made such beautiful things in this world. But You also cared to make the dirty soil that plants need. So many wonderful things happen underground to make Your world work perfectly. Thank You for making our world like this! Amen .

Wow! The soil sure is an amazing place for lots of hidden work!

Today is Green Thumb Day! Time to do Activity 57 in the Activity Book.

This Mongolian boy helps care for his family's reindeer on the tundra. Reindeer and caribou are the same animal. They are called "caribou" in North America and "reindeer" in Europe and Asia. Reindeer are often tame and used by people. Caribou are wild and migrate in large herds.

CHAPTER 20

Special Plants in Special Places: The Tundra

About the Tundra

The tundra is found in areas with very cold, long winters and very short, cool summers. Tundra doesn't usually get over 50°F (10° C) even in its warmest month. There are no trees on the tundra, and it's almost always windy! Tundra can be found in two kinds of places: The **arctic tundra** is in the far northern parts of Asia, Europe, and North America. **Alpine tundra** is found on the tops of high mountains around the world. There is a part of Antarctica that some say is like tundra. But it only has two kinds of plants and is much colder and drier than any other biome.

For behold,
He who forms mountains,
And creates the wind,
Who declares to man what his thought is,
And makes the morning darkness,
Who treads the high places of the earth—
The LORD God of hosts is His name. (Amos 4:13)

This verse reminds me of the tundra!

Arctic Tundra

The tundra in northern arctic places has a very short growing season. There are only about 60 days in each year when the temperature is above freezing. This isn't enough time for trees to build cells, so trees cannot grow on the arctic tundra. Sixty days is also not enough time for soil to thaw any deeper than a few inches. Below that, the soil is always frozen and is called **permafrost**. Only smaller plants with shallow roots can

soak into the frozen permafrost. Because the arctic tundra is flat, water doesn't drain. It stays in the ground, making the soil soggy and collecting in marshes and ponds.

Alpine Tundra

Some mountains are so tall that trees can't grow on their tops. This treeless area is called the alpine tundra.

live in the arctic tundra.

The arctic tundra only gets as much moisture as some deserts. But it isn't dry like a desert. Because the climate is so cold, very little water evaporates. In summer, the rain and melting snow cannot

Permafrost is not as common in alpine tundra as it is in arctic tundra!

If you could step into this picture (left) and start hiking toward these high mountains, you would first travel through a deciduous forest. Do you see the bright fall colors of the aspen leaves? Then, after a

Icy permafrost can be seen on the edge of this pond in the arctic tundra.

Glacial ice can crack, showing a beautiful blue color in its depths.

while, you would climb into a coniferous forest with tall dark trees. But then the trail would get steeper. There would be less air and colder temperatures the higher you went. The trees would get smaller and smaller. When the trees are about your height, there wouldn't be very many of them, and they would look windblown. Soon there would be no trees. At this point, you would be entering the alpine tundra. This level where trees stop growing is called the **timberline** or **tree line**. The first snow of the tundra's winter has already fallen on these mountains!

The alpine tundra has a longer growing season than the arctic tundra. And the alpine tundra gets more moisture. In winter, alpine tundra usually gets very deep snows which may not completely melt in the summer. When a patch of snow never melts from year to year, it's called a **glacier**. The snow on a glacier gets packed down by its own weight until it becomes ice. A glacier's weight can make it slowly slide down the mountain.

Time to do Activity 58 in the Activity Book!

Special Plants in the Tundra

The tundra is cold and usually windy. Wind often picks up tiny ice crystals and blows them against the plants. Plants living on the tundra need to be tough! God perfectly made tundra plants to live and make seeds in this harsh place.

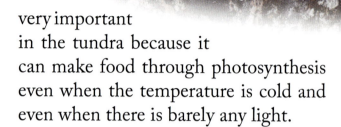

God gave some tundra plants, like this edelweiss, fuzz to protect them from the wind.

Most tundra plants are small and hug the ground where it's less windy. The most common plants are mosses, grasses, and shrubs. Most alpine plants are perennials, and some can live more than 100 years! Like flowering desert plants, tundra flowers can bloom and make seeds quickly in the short growing season.

Lichen can live in the tundra because it grows close to the ground—usually on rocks. In fact, it helps break off tiny pieces of rock to make soil. Lichen is very important in the tundra because it can make food through photosynthesis even when the temperature is cold and even when there is barely any light.

Definitions

Cushion plants grow close to the ground where the wind is calmer. Many have long taproots to reach water and nutrition in the rocky alpine tundra.

Lichen (lie-ken) can grow in most biomes, but it's one of the most common things growing in the tundra. Lichen is strange! It's not exactly a plant. It's made up of a kind of fungus and little plant-like things that do photosynthesis. The fungus protects the plant-like things and gives them minerals. The plant-like things share the sugars they make with the fungus. There is symbiosis between them!

This pale reindeer lichen is the main food for reindeer during the tundra's cold months.

Time to do Activity 59 in the Activity Book!

Yellow-bellied marmots are large rodents that live only in the Rocky Mountains of North America. They eat plants all summer long, then hibernate during the long winter. They have a loud warning chirp that can be heard very far away.

Yellow-bellied marmot

Special Animals in the Tundra

Isn't it amazing that God made some plants able to live in such cold places? Now let's learn about some of the amazing animals God put in both kinds of tundra!

Lemming

Snowy owl

Herby's List of Arctic Tundra Animals

- Lemmings are six-inch (15cm) rodents that need very little sleep. After only a small nap, they can stay awake for hours in the night. They eat plants but become food for meat-eating animals. The number of lemmings can change quickly! In one summer, a few of them can multiply to almost too many. God made lemmings able to have lots of babies because so many of them are hunted by animals or die trying to cross streams. Lemmings keep warm by burrowing under the snow and by making nest-like shelters of grass, fur, and feathers.

- Snowy owls eat mostly lemmings. If there aren't many lemmings when it's time to have babies, the owls wait to lay eggs until there is more food. Female snowy owls and their babies are brownish, but grown males are white. These owls have a thick layer of down feathers next to their skin to keep them warm. Even their legs and toes are covered with feathers!

- Arctic foxes also eat mostly lemmings. Mother foxes have up to 19 babies at a time! Those babies grow into adults in only one year. Then they can have a lot of babies of their own. Arctic foxes have thick, warm fur and a layer of fat under their skin. Fat helps keep warmth inside the animal. They often burrow in snow to keep warm.

- Reindeer (caribou) are the only kind of deer in which both males and females have antlers. Each spring, the females lose their antlers after their babies are born. Males lose their antlers in the fall. God keeps reindeer warm by giving them hollow hairs and a layer of fat under their skin. Each hollow hair traps air inside itself. This air stays warmer than the surrounding air and keeps the animal warm.

- Polar bears live where the arctic tundra meets the ocean. This is the best place to snatch their favorite food (seals) out of the water. Of all the meat eaters that live on the earth's land, polar bears are the largest. These bears look white, but each of their hairs is actually clear. The hairs are hollow, which makes them reflect light and look white. These hollow hairs and a thick layer of fat keep the bears warm as they swim in icy water. Back on land, polar bears warm up because God has given them black skin under their transparent fur to help them absorb heat from sunshine.

- More mosquitoes live in the arctic tundra than in any other biome. Adult mosquitoes lay their eggs in still water. The babies hatch and grow up underwater and fly out as adults. Since so much of the arctic tundra is ponds and marshes, there are a lot of places for huge numbers of mosquito eggs to be laid. When it's time for the adults to fly, the sky can sometimes look gray above the tundra!

Arctic fox

Polar bear

Mosquitoes

Flora's List of Alpine Tundra Animals

Each of the alpine mountain tundras around the world are home to different animals.

Mountain goats

Guanacos

- Rocky mountain goats live in the high mountains of North America. Their pure white fur grows in two layers—a short fuzzy layer near the skin and a long winter coat made of hollow hairs. To help mountain goats climb safely, God gave them sharp hooves and special foot pads for sticking to rocks. Baby mountain goats try to walk and climb right after they are born! Their mothers usually stand below them on steep parts of the mountain to keep them from falling!

- Guanacos are camel-like animals that live in the Andes Mountains of South America. Baby guanacos are called **chulengos** and can walk as soon as they are born. God gave them this ability so they can keep up with their herd. Guanacos have very thick skin on their necks to protect them from the bites of enemies.

- The chamois ("sham-wah" or sometimes "shammy") is a goat-like antelope that lives in the high mountain tundra of Europe. Their fur is rich brown in the summer and changes to light gray in winter.

• Africa doesn't have very many areas of tundra, but one of these areas is home to the ice rat. Each ice rat has its own area of plants where it doesn't allow other rats to feed. But when it's cold, all the rats come together underground and cuddle in a heap in their burrows. Each family has several burrows that are connected by tunnels.

• Asia has some of the tallest mountains in the world. The wild yak lives in these mountains—higher up than any other alpine tundra animal in the world. The wild yak is an ox-like animal that lives in herds. God keeps the yak warm with two layers of fur. The coat next to the skin is long and sometimes hangs like a skirt from the yak's belly. Yaks also have a special way of digesting their food that makes their belly hot inside and helps keep them warm. There are some yaks that are not wild. For years, people have used these tame yaks as work animals or to provide milk and meat.

Chamois

Ice rat

Yaks

Prayer

Dear God, we are so amazed that You created plants and creatures that are able to live in the cold, windy tundra. Thank You for caring for them. We praise You for the beauty of tall, steep mountains and the plants and animals living in their high places. Amen.

Time to do Activity 60 in the Activity Book!

Low plants
and lichen in
alpine tundra

UNIT 6
Fruit: Designed and Delicious

It's going to be fun learning about the fruits God designed. Some fruits are luscious and colorful. Some are dry and spiky.

But God has given them all the same purpose: to spread their seeds!

 ## Hymn to Sing

Can a Little Child Like Me

Can a little child like me
Thank the Father fittingly?
Yes, O yes! be good and true,
Patient, kind in all you do;
Love the Lord and do your part;
Learn to say with all your heart,
Father, we thank Thee!
Father, we thank Thee!
Father in heaven, we thank Thee!

For the fruit upon the tree,
For the birds that sing of thee,
For the earth in beauty dressed,
Father, mother, and the rest,
For thy precious loving care,
For thy bounty ev'rywhere,
Father, we thank Thee!
Father, we thank Thee!
Father in heaven, we thank thee!

You can listen to this hymn by searching for "Can a Little Child Like Me" on the internet.

Memory Verse

"Fear not, you beasts of the field, for the pastures of the wilderness are green;
the tree bears its fruit;
the fig tree and vine give their full yield."
(Joel 2:22 ESV)

CHAPTER 21
What is a Fruit?

Fruits Enclose the Seeds

A **fruit** is a shiny, sweet cherry that gives you its dark, delicious juice as you carefully chew around the pit. A fruit is a pear that becomes liquid in your mouth, with its fragrance filling your nose from the inside. A fruit is a creamy banana that satisfies your hunger. A fruit is a tangy peach whose juice explodes out of its fuzzy skin if you don't slurp hard enough. A fruit is even a crunchy cucumber and a salty little tomato on your salad. It's the grain in your bread and the chocolate in your dessert. A fruit is a blessing from God!

We are amazed by God's beautiful fruits. Now let's learn what a fruit is.

A **fruit** is the part of a plant that contains the seeds. It comes from the flower. A flower's ovary becomes its fruit after pollination.

Many changes happen to a flower after pollination. As the seed begins to grow, hormones are made which cause the flower's petals and stamens to fall off. A tough covering is formed on the seed to protect it. Around the seed's covering, the ovary gets larger and is called a fruit. Fruits can be soft and juicy or dry and hard. They come in many shapes and sizes.

Fruits can be delicious and very interesting to eat. We often think of sweet fruits like pineapples, oranges, bananas,

Parts of an Apple Blossom

and strawberries. But there are also many edible fruits that aren't sweet. We usually call fruits "vegetables" when they aren't sweet. Some of these non-sweet fruits are beans, peas, corn, squash, cucumbers, tomatoes, peppers, and eggplant. But these are still fruits because they contain the seeds of the plant!

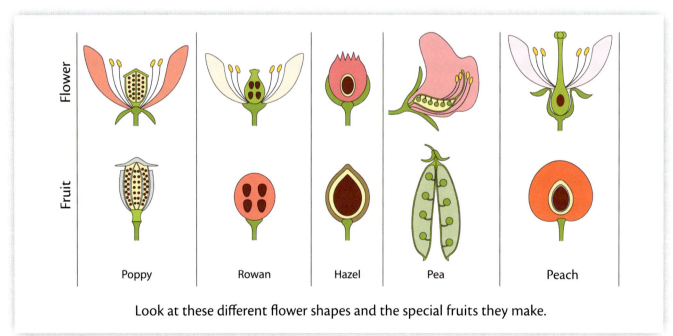

Flower	Poppy	Rowan	Hazel	Pea	Peach

Look at these different flower shapes and the special fruits they make.

Time to do Activity 61 in the Activity Book!

190

Fruits Grow and Change

When fruit begins to grow on an apple tree, the tiny fruits are hard and sour. They don't smell much like apples. We say they are **unripe**. As time goes on, we can use our mouths to tell us when the apples are no longer sour but sweet. At the same time, the apples also smell delicious because of fragrant chemicals made just under the skin. Now we know the apples are **ripe**!

But what happens if we wait too long to eat the apples? They are no longer crisp and don't taste as sweet. They are **overripe**. When the apples were ripe, they were crisp because their cells were joined tightly together. When we took a bite, the cells popped open and squirted out their sugars. But now that the apples are overripe, the cells are no longer joined tightly together. When we chew the apple, most of the cells separate from each other instead of popping. We don't taste as much sweetness. But the sweetness is still there, and God gives us a way get to it! We just need to cook the apples and make applesauce! The heat will pop the cells and make the applesauce sweet!

Ethylene is the hormone that makes fruit become ripe! It also can make fruit overripe.

In the Bible, God often uses fruit as an example to show how He will bless His people.

"I will make them and the places all around My hill a blessing; and I will cause showers to come down in their season; there shall be showers of blessing. Then the trees of the field shall yield their fruit, and the earth shall yield her increase." (Ezekiel 34:26-27)

Prayer

Lord, thank You for making so many interesting and delicious fruits! You have given us an amazing variety of flavors, colors, textures, and shapes of fruits to eat. We could never think of these different flavors ourselves. Thank You for Your blessings! Amen.

Time to do Activity 62 in the Activity Book!

Did you know that chocolate comes from the seeds of cocoa fruits like these?

Wind carries the seeds of this cotton grass in the North American tundra.

CHAPTER 22

The Purposes of Fruit

"[God] satisfies your mouth with good things." (Psalm 103:5)

This Bible verse shows that God blesses us with tasty food!

God Made Fruit for Us to Enjoy

God made people. He meant us to need food to stay alive. What if God made all the nutrition we would need, mixed it all together, and gave it to us in pills. If He did that, we would have to swallow these same pills every day. There wouldn't be much to enjoy about eating that way!

Instead, God shows His loving care for us by giving us many different foods with many different flavors. And He gave us creative minds to think of recipes for combining foods into even more yummy things to eat.

We couldn't enjoy all these tastes if God didn't give us the ability to sense them. Our tongues can sense the basic tastes of sweet, sour, salty, bitter, and savory. Our noses make these tastes more complicated and wonderful by noticing tiny amounts of chemical combinations as we chew our food. God made our senses of taste and smell work together with delicious food so we can enjoy His gifts!

Fruits give our tongue several kinds of sugar for sweetness. Our tongue also senses fruit acids for a pleasant tartness. Then, the inside of our nose picks up the smell of fruit from inside our mouth. Our brain receives this sensed information about the fruit from the nerves of the mouth and nose. Then we can thank God for satisfying our mouth with good things!

Pruning an orchard

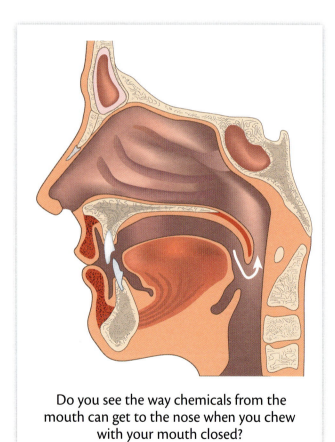
Do you see the way chemicals from the mouth can get to the nose when you chew with your mouth closed?

Farmers with fruit orchards can improve the flavor of fruit. They can prune away extra branches because the taste and color of fruits improve when more sunlight reaches them. Farmers can also pick some of the fruit early so that more nutrition goes to the remaining fruits. And they can make sure not to overwater because overwatering can weaken the fruits' flavor.

Thinning apples

Time to do Activity 64 in the Activity Book!

Bye-bye, Seeds

God has put many kinds of fruit in the world for us to enjoy. But another purpose of fruit is to spread seeds!

Grown up seeds can't stay inside the plant. They must get to the soil! If plants only dropped their seeds onto the ground below them, the seeds would be crowded. After they sprouted, all those new plants would quickly use up the nutrition and water in the soil and have a hard time getting enough sunlight. They would become weak, and they would make the mother plant above them weak too. It would be better if plants had ways for their seeds to be spread farther. They do!

God has designed many different ways for fruits to help spread out their plant's seeds:

- God made many fruits tasty, nutritious, and brightly colored so that animals would be attracted to them. Animals

eat the lovely fruits and end up eating the seeds too. The seeds pass through the animal's body and come out later in the animal's waste. Since animals move around a lot, the seeds now have a new place to grow.

- Some animals bury nuts in different places so they can eat them later. But they often forget about some of their nuts. These forgotten nuts sprout into trees. This means that the animals have planted some nut trees!

- A tall, grass-like plant in South Africa tricks dung beetles into planting its seeds! Dung beetles normally bury animal waste (dung) and lay their eggs in it so the hatched beetle babies have something to eat. But God made the seeds of this plant look like the dung of the antelope that live in the same area. The seeds even have the same stink as the dung! The odor keeps the seeds from being eaten by

Dung beetle and dung

This hazelnut was buried by a squirrel and then forgotten.

the plants flip out the seeds, and the beetles rush to bury them.

- Some fruits shoot their seeds away from the plant! This might happen as the fruits dry up. Others break open and shoot seeds when they are touched. And some fruits swell up with so much water that they pop open.

- Some plants use burrs to help spread their seeds. A **burr** is a rough or prickly envelope of a fruit. The prickles on the fruit help it stick to animals and people so it can be carried off to spread its seeds.

- God designed some seeds to be carried away from their plant by the wind. Some fruits have fluffy hairs that catch the wind and carry their seeds away.

small animals, but it also fools the dung beetles into thinking the seeds are dung. On the plant, the seeds are held in cup-like fruits until the dung beetles are most active. Then

If you let the okra fruits in your vegetable garden dry up, they will split open and shoot their seeds when you touch them!

Hi! I'm Puff. I love blowing God's flying seeds to the places He wants them to grow!

The pumpkin-shaped fruits of the sandbox tree explode with the sound of a firecracker! The fast-flying seeds can injure people! Look at the bark to see why this tree is sometimes called the monkey-no-climb tree.

Clematis going to seed

Water carries floating coconuts to new places. Scientists say a coconut is a fruit, a nut, and a seed all at the same time!

Boy experimenting with a tumbleweed

- Tumbleweeds are round plants that spread their seeds as the wind rolls them along.

- Some trees have wing-like fruits that send their seeds whirling through the air.

- Plants that live in or near water often have seeds that float.

[Jesus said,] "By this My Father is glorified, that you bear much fruit." (John 15:8)

Prayer

Father, we thank You for making fruits that taste so good. We really like cold watermelon, warm apple pie, and hot chocolate. And we praise You for giving plants so many ways to spread seeds. You did this so the earth would bear much fruit. We pray that You will help us understand how we can bear much fruit for You. We want You to be glorified as this verse says! Amen.

Time to do Activity 65 in the Activity Book!

God had such great purposes for designing all the different fruits!

Today is Green Thumb Day! Time to do Activity 66 in the Activity Book.

CHAPTER 23

Fruit in Its Many Forms

Oh, taste and see that the LORD is good;
Blessed is the man who trusts in Him!
(Psalm 34:8)

God made each kind of plant with a different kind of fruit. Even though there are many different fruits, we can put them into groups by looking at ways certain ones are alike.

Fruits are often separated into two large groups: fleshy fruits and dry fruits.

Fleshly Fruits

Fleshy fruits have three layers around their seeds. The middle layer is thick, soft, and often juicy. Fleshy fruits stay moist even after their seeds are completely formed.

There are many kinds of beautiful, fleshy fruits. The fleshy fruits can be divided into three smaller groups.

The Three Layers of Fleshy Fruits

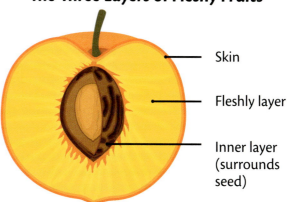

Skin

Fleshly layer

Inner layer (surrounds seed)

Flora's List of Fleshy Fruits

- **Drupes**, or stone fruits, have one seed. That seed is surrounded by a hard, stony covering called a **pit**. The pit is not part of the seed. It's the inner layer of the fruit. Outside the stony pit is a fleshy layer that's soft and often juicy when ripe. Peaches, plums, cherries, apricots, and olives are drupes.

- **Berries** don't form a stony pit around their seeds. Their inner layer is softer. The fleshy layer of berries is divided into sections that contain several or many seeds. Berries often have thin skin. Blueberries, tomatoes, peppers, and grapes are berries. Citrus fruits, like lemons and oranges, are a type of berry with tough rinds instead of thin skin. Pumpkins and gourds are another kind of berry with hard rinds for skin.

- **Pomes** are interesting fruits whose juicy, fleshy part is not made from the ovary. Instead, it's made from the stem that held the flower. The ovary of the flower becomes the core around the seeds. The stem that held the flower swells to become the part of the fruit we like to bite. Apples and pears are pomes.

Fleshy Fruits

Drupe   Berry   Pome

It's amazing to see how God designed the fruits in these groups! But have you noticed that there are many fleshy fruits that don't seem to fit into these three groups? God is so creative that He made many fruits that surprise us:

Blackberries are called berries in the grocery store. But in science, they are not called berries. One blackberry flower actually has a lot of ovaries that become a lot of drupes. Each tiny ovary makes a dark, juicy fruit with a single seed inside. That means that one "blackberry" is really a collection of tiny drupes!

Strawberries fool us because the red part isn't really a fruit! As we bite into it, we are actually tasting the place where the flower was attached to the stem and swelled into deliciousness. Lots of tiny brown fruits, containing one seed each, are attached to the outside of the red part. We feel them as little crunchy things between our teeth! So the fruits on a strawberry are not fleshy, but dry!

Pineapples look so prickly on the outside, but when ripe and sweet, they are luscious inside! Pineapples are formed by a whole bunch of flowers blooming together in a cluster. The flowers' ovaries grow together, which means

Blooming pineapple

their fruits will be joined into one large thing to eat. A pineapple is actually a whole bunch of berries joined together!

Time to do Activity 67 in the Activity Book!

Dry Fruits

Dry fruits have only one layer covering their seeds. The covering is leathery, not fleshy. It dries up when the seeds are grown. Let's look at a few of the groups scientists have made for dry fruits.

Herby's List of Dry Fruits

- **Pods** are fruits whose ovary has only one section, but there are a number of seeds in it. Beans and peas have pods.

Peanuts are not nuts! Peanut fruits are pods which develop underground.

- **Capsules** are fruits whose ovary had many sections, each with many seeds.

- **Grains** are fruits whose

Poppy seeds form in capsules.

ovary wall is connected to the single seed.

- **Achenes** (uh-keens) are fruits whose ovary is not connected to the single seed inside. That's why a sunflower seed rattles if you shake it. Their coverings are not as hard and woody as nut coverings. Achenes are a complicated group. There are many kinds of achenes that look very different from each other. Some of the plants that have achenes are sunflowers, maples, dandelions, and strawberries.

Sunflower achene with its seed inside

- **Nuts** are fruits like achenes, but the ovary is a hard, woody shell around its one seed.

Dry Fruits

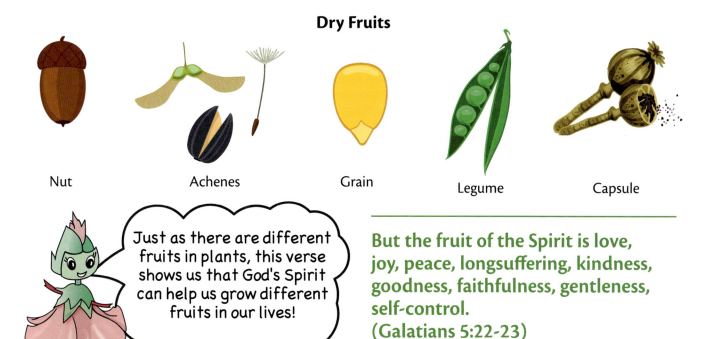

| Nut | Achenes | Grain | Legume | Capsule |

Just as there are different fruits in plants, this verse shows us that God's Spirit can help us grow different fruits in our lives!

But the fruit of the Spirit is love, joy, peace, longsuffering, kindness, goodness, faithfulness, gentleness, self-control.
(Galatians 5:22-23)

Prayer

Dear God, we pray that, when we eat each day, we will be thankful for the many wonderful and delicious fruits You have blessed us with. It's amazing that You have designed so many ways for fruits to spread seeds! Amen.

Time to do Activity 68 in the Activity Book!

Hasn't God designed such interesting fruits? They all have the purpose of spreading a plant's seeds. But He blesses us by making fruits taste good too!

Today is Green Thumb Day! Time to do Activity 69 in the Activity Book.

Nuts, seeds, and fleshy fruits make delicious healthy toppings!

Up to 150 banana fruits can grow at the same time on one banana plant. This tropical rainforest plant isn't a tree; it's an herb!

CHAPTER 24

Special Plants in Special Places:
The Tropical Rainforest

About Tropical Rainforests

Tropical rainforests are found near the **equator**. The equator is the widest part of the earth, halfway between the north pole and the south pole. Because the equator is at the middle of the earth, the sun shines directly on it all year long—there are no seasons there. This makes temperatures warm or hot all the time.

This picture is showing summertime for the *top* part of the world. Notice how the daytime sun is shining directly on the top part of the earth. It's also shining directly on the equator. In six months, it will be summer for the *bottom* half of the world. The earth will have

moved around to the other side of the sun. The earth's tilt doesn't change as it moves around the sun, so now the sun will shine directly on the bottom part of the earth. It will still be shining directly on the equator too. This means that the equator will have warm temperatures all year long. Warm temperatures are the perfect place for tropical rainforests.

Earth's Seasons

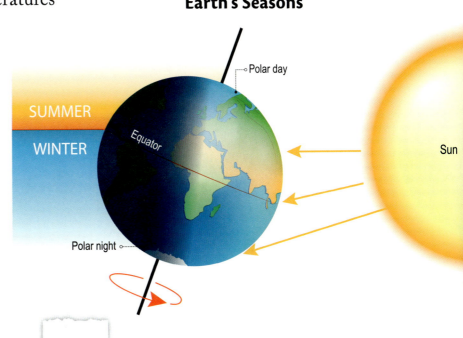

211

Tropical rainforests get more moisture than any other biome. If you lived in a tropical rainforest, you would have a thunderstorm almost every day!

Some months are wetter than others, but even the driest month of the year gets more rain than many deserts get in a whole year!

Because of its warmth and wetness, the tropical rainforest has more kinds of plants than any other biome—about three million! And if you were able to weigh all the plants in the biomes, you would find that the tropical rainforest also has the greatest weight of plants.

Time to do Activity 70 in the Activity Book!

Special Plants in Tropical Rainforests

Tropical rainforests are filled with evergreen plants. But these evergreens are not the evergreens that make cones and have needles. Tropical evergreens have broad leaves. Their leaves don't usually fall off all at once like the broad leaves of deciduous trees. They fall off one at a time all year long.

Plants in the tropical rainforest grow in layers. The **canopy** layer is so crowded with leaves that it blocks wind and most sunlight from the forest beneath it. Since there is little wind in this layer, there are no seeds that need to be spread by the wind. Instead, seeds grow inside fruits that are brightly colored so animals can see them. Hungry birds, mammals, and reptiles enjoy the fruits and spread the seeds through their waste.

The tropical canopy is full of interesting plants that spend their whole lives attached to trees, with their roots never touching the ground. Orchids, mosses, and ferns live on trees. Their roots get water directly

Birds eating a ripe papaya. Figs, mangos, and papayas are just a few fruits in the tropical canopy.

This branch in the canopy is home to orchids and ferns.

emerge) from the canopy. This is called the **emergent** layer. There's no shelter from sun and wind in this high layer. Many of these trees have seeds that are spread by wind instead of by animals.

The **understory** layer of the rainforest is more open than the canopy. It is mostly made up of bare tree trunks and vines that climb up on the trees to reach sunlight. Once these vines grow high into the other layers, they reach from tree to tree, making bridges for creatures to travel on. The shorter plants that grow in the understory have big leaves. Big leaves help absorb as much sunlight as possible in this dim layer. God has made the flowers of the understory large and

from rain and from moisture in the air. Instead of getting nutrition from soil, they get it from rotting leaves that fall into crevices on the trees.

Some very tall trees push through (or

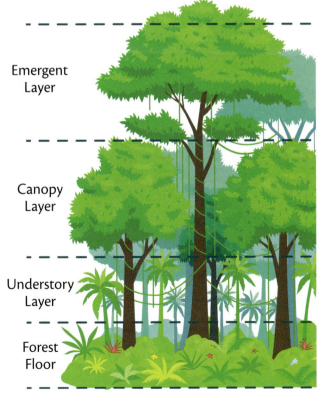

Emergent
Layer

Canopy
Layer

Understory
Layer

Forest
Floor

Forest Layers

This canopy-dwelling plant collects rainwater and gives tree frogs and mosquitoes a place to lay their eggs.

These tropical emergent trees have winged seeds. This kind of tree is a favorite place for bees to build hives.

fragrant to help pollinators find them. The understory is the most humid layer in the rainforest.

The **forest floor** is very shaded and dark. You might need a flashlight to see an ant on the ground. Not many plants grow at ground level because there isn't enough sunlight. Leaves and fruits fall to the ground and are turned into soil quickly because of warm temperatures, moist air, and lots of soil creatures. But the soil isn't very nutritious in tropical rainforests. The nutrition that's made by rotting leaves and fruits is quickly used up by the large number of plants growing there. Also, the abundant rain washes through the soil and carries away its nutrition.

A **liana** (lee-ah-nuh) is a woody vine that climbs up trees to reach sunlight.

214

This silk cotton tree uses fuzz to catch wind and carry away its lightweight seeds. The fuzz repels water and was once used as stuffing in life jackets to help them float.

Brazil nut fruits are large and heavy enough (5 pounds, or 2 kg) to fall from the emergent layer straight to the rainforest floor. There, the nuts are eaten by the only animal able to crack them open—the agouti (uh-goo-tee), a large rodent with chisel-like teeth. The hard fruit has one hole in it which allows an agouti to grab it in its mouth and carry it away. Some of the seeds will be buried (and forgotten) by the agouti. The seeds will wait in the ground a very long time until a nearby tree falls, which gives them enough light to sprout and begin growing. Once it sprouts, a Brazil nut tree can live about 1000 years!

Time to do Activity 71 in the Activity Book!

Special Animals in Tropical Rainforests

Each of the layers in a tropical rainforest is a different neighborhood for different animals! Let's look at some of the animals in each tropical forest layer. God made each one just right for its home!

Many tropical rainforest animals live in one layer but find food in another.

The   trees of the LORD are full of sap ...
Where the birds make their nests.
(Psalm 104:16-17)

Emergent-Layer Animals

Animals living in the emergent layer aren't very big. The top branches of trees are too thin to hold heavy animals. Many animals in the emergent neighborhood travel by flying or gliding between the tree tops.

The harpy eagle weighs only 20 pounds (9 kg), but its feet are almost as big as a man's feet. These eagles live in the treetops of the emergent layer, watching the canopy below for prey. Harpy eagles kill medium-sized animals with their

Scarlet macaws

strong feet and grizzly bear-sized claws. They eat on the ground, often taking several days to consume an animal even if the meat gets rotten. Their favorite foods are mammals: sloths, monkeys, and opossums.

Scarlet macaws are bright, colorful, noisy birds that live and fly in the emergent layer. They often eat in the canopy where they find fruits, nuts, nectar, and insects.

Harpy eagle

This baby capuchin (kah-poo-chin) monkey is so small and light that he can stand near the end of tree branches. Capuchin monkeys are some of the smartest monkeys God made. They are good at figuring things out. They spread out nuts to dry and then travel to a faraway streambed to find rocks they can use to crack open the nuts. During mosquito season, capuchin monkeys squish millipedes in their hands and rub them on their backs to repel mosquitos!

Blue morpho butterflies are so large and so bright that low-flying airplane pilots often see them flitting in the emergent layer.

Capuchin monkey

Blue morpho butterflies

Canopy-Layer Animals

The canopy layer is the neighborhood with the most animals. It produces the most food and has the best hiding places. Because it's hard to see through the crowded leaves, many animals use loud noises to find their family and friends.

Howler monkeys have one of the loudest calls of any land animal. Their tails grab branches and vines and help them balance as they travel in the canopy trees.

Toucans (too-cans) are noisy birds! They make many different sounds with their voices and bills. Because of their large bills, toucans don't fly easily. In fact, they spend more time hopping than flying.

God made many insects able to hide in plain sight with **mimicry**. This mantis hopes his next meal will think he's an orchid flower and will come near enough to be grabbed!

Red-eyed tree frogs like to hide on the underside of leaves. With their feet tucked in and their eyes closed, their green skin matches the leaves and helps them hide. If a hungry animal discovers him, the little frog opens his red eyes, spreads out his orange toes, and surprises the

My favorite insect is the orchid mantis. I'm so happy God made them!

enemy! The frog hopes this will give him a little more time to escape.

Draco lizards have flaps of skin that spread out as they glide between trees.

Understory-Layer Animals

The understory layer of rainforests is very dim. Animals that live in this neighborhood are usually **camouflaged** so they will be hard to see in the deep shade. Predators in the understory spend their time up off the ground, waiting in small trees where they can see the forest floor below. When a tasty animal walks underneath, the predator can drop down on top of it.

Boa constrictors wrap themselves around their prey and squeeze it to death before they eat it.

The understory has room for bats to fly and hunt insects. God has given this common big-eared bat an amazing ability to echolocate, even in the noisy,

Definitions

Mimicry is when one living thing looks like a different kind of living thing.

Camouflage is when a creature's appearance matches its surroundings and helps it hide.

night-time rainforest.

The postman butterfly has plenty of room to fly through the understory. Its caterpillars feed on passion flower leaves, which makes the caterpillars and adults stinky. It also makes them poisonous. The butterfly's bright colors warn predators that it's not safe to eat.

Jaguars and leopards are both large cats that live in the rainforest understory. Jaguars (top) live in Central and South America. Their ring-like spots have small dots in the middle. Leopards live across the ocean in Africa and Asia. A leopard's ring-like spots have no dots inside.

Forest Floor Layer Animals

The forest floor is full of animal life, especially insects and spiders. This neighborhood is also home to large animals that can't climb trees, and to some that can.

Millions of leafcutter ants live in colonies in the forest floor. Using more than 1,000 underground rooms, the colony grows enough fungus to feed their babies. How do they make the

fungus? When a new ant colony is started, a queen ant brings along a piece of fungus from the old colony. Other ants start collecting pieces of leaves. Long

lines of leafcutter ants with their leaves can be seen traveling on vines and trees all over the rainforest understory. They look like a parade of marching ants carrying green flags! Then, back at the colony, different ants chew up the leaves, "plant" bits of the fungus on it and wait for their harvest to grow!

Of course, with all those ants, we aren't surprised that God made anteaters that live on the forest floor! The giant anteater has long, sharp claws it uses to tear up ant nests. Then, its two-foot-long tongue flicks in and out of the nest. The ants are easily caught in the sticky saliva that coats the tongue. When the anteater isn't using its claws, it curls them back to keep them from wearing out as it walks.

Tapirs are pig-shaped mammals that live on the forest floor. They like spending time in and under water looking for plants to eat and keeping cool. Tapirs have long, bendy noses that help them reach tasty leaves. Tapir babies have stripes for camouflage. A group of tapirs is called a *candle*.

Gorillas are mammals that use their knuckles to walk on all fours. They sleep in nests they make, usually on the ground. Though gorillas spend most of their time on the forest floor, they can climb trees. A group of gorillas is called a *troop*.

221

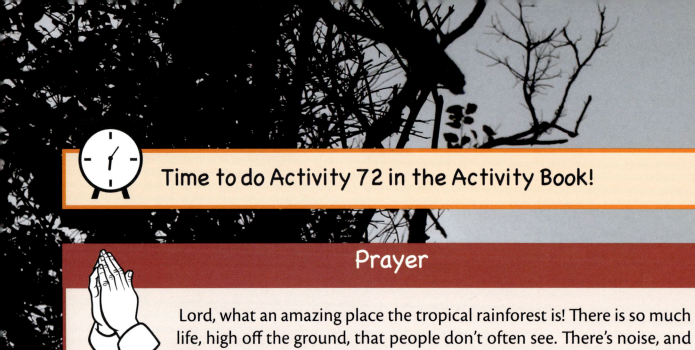

Time to do Activity 72 in the Activity Book!

Prayer

Lord, what an amazing place the tropical rainforest is! There is so much life, high off the ground, that people don't often see. There's noise, and color, and flavor. Thank You that we can learn about tropical plants and animals. Thank You that we can eat tropical foods even if we live far away from the rainforest. Amen.

Dusky langur
jumping
between trees

Mustard seed and blooming mustard plants

Seeds: A Promise of Life

A mustard seed knows it will grow into a mustard plant. And it knows how to do it! God has put those instructions inside each seed, and those instructions can't fail!

I like our new memory verse. God uses the mustard seed to teach us about faith!

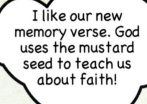

Hymn to Sing

Praise for God's Bounty (from Psalm 65)

To bless the earth, You give us,
From Your abundant store,
The waters of the springtime,
Enriching Earth once more.
The seed by You provided
Is sown o'er hill and plain,
And You, with gentle showers,
Do bless the sprouting grain.

You crown the year with goodness,
The earth Your mercy fills,
The wilderness is fruitful,
And joyful are the hills;
With corn the vales are covered,
The flocks in pastures graze;
All nature joins in singing
A joyful song of praise.

This hymn may be sung to the tune of "Stand Up, Stand Up for Jesus," which can be found on the internet.

Seeds are wonderful little packages of promises. Each seed has a baby plant inside it that's all set to live, grow, and make more seeds.

Memory Verse

"If you have faith as a mustard seed, . . . nothing will be impossible for you." (Matthew 17:20)

A walnut tree grows from one walnut. But when that single walnut grows into a tree, it makes thousands more! By the time the tree is 30 years old, it can make about 5,000 walnuts every other year, and about 1,000 walnuts in the years in between.

CHAPTER 25
Spectacular Seeds

Inside a seed, there's an explosion waiting to happen. The tiny seed that looks so dry and dead will burst and grow into something so much larger! It's a slow-motion explosion into something big, green, and full of life!

Seeds Hold the Promise of Life

Seeds are wonderful little packages that hold amazing promises. Let's look at the promises of seeds.

1. A seed holds the promise of becoming a certain kind of plant.

Do you remember that a flower can only be pollinated with pollen from the same kind of flower? In fact, the stigma ignores pollen from a different kind of flower. The reason God made flowers this way is to avoid confusion. A flower's ovules have very detailed instructions for what its next plant will become. But it has only half of the instructions. The other half of the instructions are found in pollen. The pollen's instructions are also very detailed. For these two halves to join together into complete instructions, they must come from the same kind of plant. If information from two different kinds of flowers could be joined, the plants that came from their seeds wouldn't look right! Plant life would be confusing, and the Bible says that God doesn't create confusion (1 Cor. 14:33).

When the instructions are joined and complete, they become a long, twisted molecule we call **DNA**. DNA stands for two words—one very long and one very short: deoxyribonucleic acid. But we'll just call it DNA. DNA tells the plant how to make the proteins it needs to become what God intends it to be.

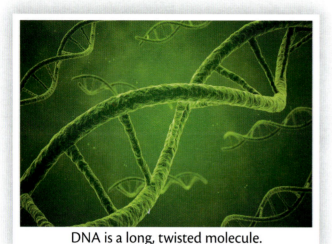

DNA is a long, twisted molecule.

And the earth brought forth grass, the herb that yields seed according to its kind, and the tree that yields fruit, whose seed is in itself according to its kind. And God saw that it was good. (Genesis 1:12)

When the instructions from the pollen join the instructions in the ovule, the ovule begins to change into a seed. This is how they fulfill God's promise that seeds would become plants "according to their kind."

Nothing would happen without DNA's instructions!

DNA has different instructions at different places along its chain. One place on the chain might be important for making green machines. Another place might have instructions for making xylem. God made these kinds of instructions permanent. But there are other places on the DNA that give instructions for things like the color of a sweet pea flower or whether a corn plant will have yellow, purple, or white kernels on its ears. God gives variety in these things by allowing parts of the DNA's instructions to be able to change with pollination. Pollen from a different plant brings in new possibilities. Because of pollination, a plant may look a little different, taste a little different, or be able to fight pests and diseases a little better than its parents. But it still is the same kind of plant.

Sweet pea flowers

2. A seed holds the promise of future generations.

When an anther makes its pollen, and when an ovule makes its set of instructions, they are making information that will be passed along from their seed to the plant that will grow from it. Every cell in the new plant will have a copy of the same DNA their seed started with.

When the time comes for the new plant to make seeds of its own, its flowers already have their half of the DNA waiting for the other half to be brought by a pollinator!

A plant comes from one seed. But God designed each plant to make a lot of seeds. Many of these seeds will grow up into plants which will also make many seeds. This has happened ever since God first created plants. And it will continue to happen because He designed plants to fill the earth!

Coconut seeds with their DNA can spread to new lands when their seeds float away from their seaside home and wash up onto another shore. God made coconut seeds refuse to sprout when they are in salty seawater where they can't live. But once on land, rains wash away the salt, and they can sprout! Coconuts are the largest seeds in the world. One tree can grow 30-75 coconuts a year.

You have DNA too! Your DNA holds the instructions to make you look the way you do. God uses DNA for all life!

God has filled much of the world with coconuts!

3. God promises to supply seeds to the earth:

God provided seeds on Earth in the past. We see God's promise to supply seeds (Gen. 1:11) whenever we bite open a fruit. The seeds are there, just where He said they would be when He created them! Isn't it amazing that today's seeds have the DNA that's been passed from plant to seed and from seed to plant over many **generations** since creation?

God is still providing seeds now. The Bible says that God is the One who provides farmers with seeds. God wants people to plant seeds and grow food. Getting food from farming is a better use of time than roaming in the wilderness, looking for food that has

Tractor sowing seeds today by scattering

field, scattering seeds with their hands. Planting seeds, especially by scattering them on the ground, is called **sowing**.

Now may He who supplies seed to the sower, and bread for food, supply and multiply the seed you have sown and increase the fruits of your righteousness. (2 Corinthians 9:10)

grown by itself. God made the seeds for food crops easy to gather and save. Today, farmers use tractors to help plant seeds. But long ago, farmers planted seeds by walking up and down the

In this verse, God is called the One who supplies seed to the sower. Since God's Word is always true, God is supplying us with seed right now! And it says He can help us do good things for Him.

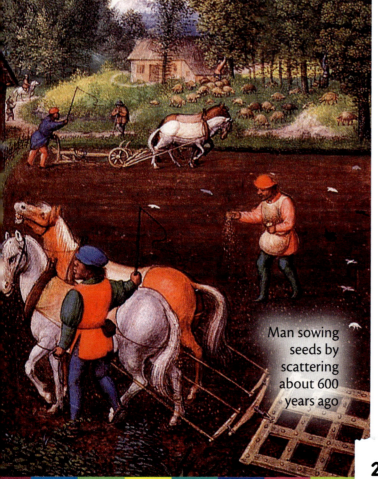
Man sowing seeds by scattering about 600 years ago

God will continue to provide seeds in the future! God has always given seeds to the earth, and He is still supplying them today. But how do we know that plants will continue to make seeds? How do we know that plants will keep growing and filling the earth with food?

At the time of Noah, the world was full of wickedness. God sent a flood over the whole world. The flood destroyed everyone but Noah and his family. It also destroyed most animals and plants. After the flood, when Noah and his family came safely out of the ark, God promised:

"[I will never] again destroy every
living thing as I have done.
While the earth remains,
Seedtime and harvest,
Cold and heat,
Winter and summer,
And day and night
Shall not cease."
(Genesis 8:21-22)

Did you notice that God said *seedtime* and *harvest* would not cease? This means there will always be seeds to plant and food to harvest on Earth!

Definitions

Sowing is planting seeds, often by scattering them.
A **generation** is a single step as living things have babies (or grow seeds) and those babies have babies. Your grandparents, your parents, you, and your children are four different generations.

Time to do Activity 73 in the Activity Book!

Orchids have very small seeds that are carried away by the wind. One seed pod can contain three million seeds!

Flora's Fun Facts About Fantastic Seeds

- A seed is an ovule with a covering that protects it. A seed contains a complete baby plant and some food to help it grow.

- Plants, like grasses and creepers, often make more plants by sprouting new ones out of their sideways stems. The DNA of these new plants is exactly the same as the parent plant. The new plants will have the same problems as the parent plants. But when new plants are made from seeds, the DNA is a little different from the parents' DNA. This means it's possible for the new plants to be a little different or better than the parent plants in some ways.

- Most of the food we eat from plants comes from plants that have flowers and make seeds. But not all plants make flowers and seeds. Ferns and mosses don't make flowers or seeds. Conifers make seeds but no flowers. We don't eat many ferns, mosses, or conifer trees.

- Seeds in the ground don't sprout until there is a certain amount of moisture. God makes seeds wait to sprout until their plant will have enough water to grow.

- Seeds dry up as they grow inside their fruits. If they stayed moist, they might leave their plant and sprout before there is enough moisture in the soil.

- When living things dry up, certain important molecules are destroyed. How can seeds still grow into live plants if they have dried up? God protects their molecules by using sugar to replace the water as it dries up. The sugar might be thick like honey or, in very dry places, it might be like hard candy, but it protects the seed's life. Then, when rain falls on the seeds, the sugar is washed away and the seed can sprout!

- The smallest seeds on Earth are as small as dust.

Prayer

Thank You, Father, for making a way for plants to fill the earth. Seeds have so much information, goodness, and beauty waiting to explode. Seeds look so small and simple, but they carry a promise for the future inside them. Please give us the faith of a mustard seed! Amen.

Time to do Activity 74 in the Activity Book!

CHAPTER 26

An Ovule Becomes a Seed

God Makes Variety

God is so wise! He made the DNA of plants in such a way that each plant will always be told to produce plants of the same kind—a corn plant makes seeds that will always become more corn plants. But God is also creative! He gave some parts of the DNA the ability to give **variety** to new generations of plants.

Corn plants have variety. They can grow kernels of various colors. A kernel's color is one of its **traits**. Kernel size and smoothness might be other traits of corn. Let's learn how variety happens in plants!

Inside a seed, there's an explosion waiting to happen. The tiny seed that looks so dry and dead will burst and grow into something so much larger! It's a slow-motion explosion into something big, green, and full of life!

Each of the corn kernels below is a seed. Each seed was pollinated separately with a different piece of pollen. Some pieces of pollen helped their seeds become purple; other pieces of pollen helped their seeds become yellow.

In growing colorful corn, four different things can happen when the pollen's instructions meet the ovule's instructions:

But what happens when the instructions they give are not the same? What if one says, "Purple," and the other says, "Yellow"? Amazingly, the new kernel will always be purple! God has designed purple to be **dominant**. Think of dominance as though purple is selfish and always gets its way.

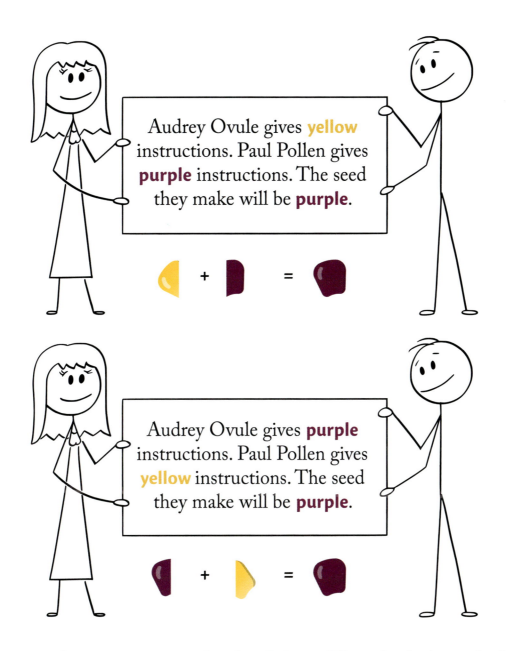

Audrey Ovule gives **yellow** instructions. Paul Pollen gives **purple** instructions. The seed they make will be **purple**.

Audrey Ovule gives **purple** instructions. Paul Pollen gives **yellow** instructions. The seed they make will be **purple**.

Scientists study variety in many kinds of plants. They think about the DNA and the instructions that the ovules and pollen give. They want to know the dominant traits. By understanding these things, they can grow plants with certain qualities. The study of DNA and traits is called **genetics**.

Time to do Activity 76 in the Activity Book!

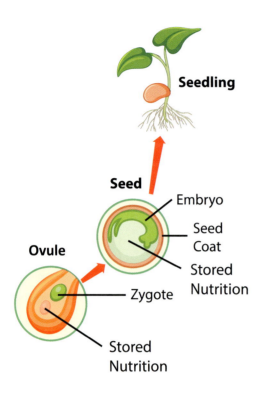

Seedling

Seed

Embryo

Seed Coat

Stored Nutrition

Ovule

Zygote

Stored Nutrition

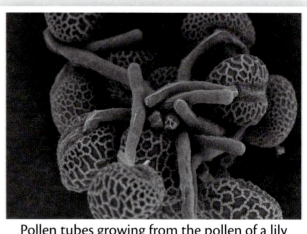

Pollen tubes growing from the pollen of a lily

From Pollination to Grown-Up Seed

Let's look closely at the way God makes a seed grow from an ovule.

1. Pollen lands on the flower's stigma and makes a pollen tube down to the ovule.

2. The pollen's instructions travel down the pollen tube.

3. The pollen's instructions enter the ovule.

4. The pollen's instructions join the ovule's instructions. This becomes the beginning of a baby plant called a **zygote**.

5. The zygote starts as one cell, but soon divides into two. Its cells keep dividing until they become early roots, stems, and leaves. It is now called an **embryo**.

6. The pollen's instructions also help the ovule make a place for nutrition to be stored. This is the storage place for proteins, oils, and other things the new plant will need right after it sprouts from the seed.

Inside of a ginkgo seed showing embryo, nutrition storage, and seed coat

7. The **seed coat** hardens to protect the embryo and stored nutrition while it waits to sprout.

Prayer

Thank You, dear God, for using DNA to make sure each kind of life always remains its own kind over the generations. And thank You for allowing variety in DNA so that plants can become more helpful or interesting. It's amazing that so much happens in a little ovule to form a seed that can grow into an entire plant. We praise You for this amazing process that fills the earth with plants! Amen.

Time to do Activity 77 in the Activity Book!

CHAPTER 27

From Brown Seed to Green Plant

Do you remember that a new plant begins life as an embryo inside an ovule? Then something strange happens. The embryo dies for a while in the seed. But the seed holds a promise of a new life. Now we will learn about the **germination** of a seed into its beautiful, wonderful new life as a plant.

up before it leaves its home in the fruit. It dries up, turns brown, and dies. And it really is dead! The embryo and the area that stores nutrition has stopped growing. Nothing else is growing inside the seed. No chemical processes are happening. The seed needs no water and no nutrition. It does nothing at all for a very long time. The seed is waiting— maybe a few months, maybe a thousand years—to live again as something very different!

Jesus said to her, "I am the resurrection and the life. He who believes in Me, though he may die, he shall live." (John 11:25)

Those who believe in Jesus are like plant embryos now. One day everyone will die like seeds do, but those who believe in Jesus have His promise that they will live again!

"Most assuredly, I say to you, unless a grain of wheat falls into the ground and dies, it remains alone; but if it dies, it produces much grain." (John 12:24)

A Time to Be Dormant

We have learned that a seed dries

Jesus talked about Himself as a seed. He taught that He would die like a seed, but would live again and bring many to Himself!

Definitions

To **germinate** means to begin to grow.
A seed is **dormant** when it is being prevented from germinating.

In this verse, Jesus says that seeds need to die in order to live and grow into plants later. It's only by being dead that seeds will know the best time to come to life. They don't want to sprout at a time when there won't be enough water or when the temperature is too cold or too hot. How do seeds know when to germinate?

Since seeds are dry, they will notice when they get wet. If a seed becomes wet enough, it knows the ground has enough moisture to be a home where it can live and grow. God made seeds wait to sprout until the ground has the right amount of moisture to support growing plants. A plant will die if it gets too cold or too hot. But a seed will last through dangerous temperatures and protect its promise of life. Then, with water and friendly tem-

See if you can find the dry embryo inside this peanut.

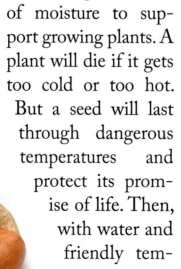

An avocado seed is always ready to sprout.

peratures, the processes of life can begin again!

Most plants have seeds that can sprout anytime there is enough warmth and water. The seeds might have to wait until spring for this to happen, but they are always ready. The seeds of tropical rainforest plants are always ready to sprout because they live in a place that is always warm, wet, and friendly to germination.

God has given other plants extra protection by requiring their seeds to have a time of waiting before sprouting, no matter how perfect the weather is.

During this time, we say the seed is **dormant**. Here are some interesting things about a seed's dormant time:

- Some seeds, like apple seeds, must be cold for a certain amount of time before they sprout. There might be a warm spell in the middle of winter, but that doesn't fool these seeds! God has given them this need for a cold, dormant time so they can make sure winter is over before they sprout into new little plants. Otherwise, they might sprout too early and freeze.

- Some plants make seeds that will sprout at different times. That way, if all the growing plants get eaten, burned, or frozen, more seeds are waiting to sprout another time.

- The hormone abscisic acid keeps seeds dormant. But once rainwater washes it away, the seeds can sprout.

- Water causes gibberellin hormones to "wake up" the seed.

- Dormant seeds can withstand temperatures far below freezing or near boiling and still be able to sprout later.

- God uses thick seed coats on some seeds to make sure they won't sprout too soon. But some of these seed coats need help to open so they can let moisture in. They need to be scratched or broken. Help comes from blowing sand, from the freezing and thawing of ice in the winter, from fire, or by passing through an animal's digestive tract.

- Some small seeds, like lettuce seeds, need to sprout near the top of the soil because they don't have enough stored energy to sprout through deep soil. These seeds know when to sprout because the stronger sunlight of spring gives them the signal. The new sprout can feed itself right away by using photosynthesis.

The fire that passed through this area helped seeds to absorb water by burning the outside of their seed coats.

Time to do Activity 79 in the Activity Book!

A Time to Grow

To everything there is a season,
A time for every purpose under
heaven:
A time to be born,
And a time to die;
A time to plant,
And a time to pluck what is planted.
(Ecclesiastes 3:1-2)

Lentils are dicot seeds.

Now let's learn about the germination of seeds: it's time to grow!

A seed is usually buried in the ground before it sprouts. Then, as the baby plant comes out of the seed coat, it immediately stretches upward. It needs to leave the darkness behind and reach the sunlight so it can make food for itself. But how can it do all this growing in the dark without the energy that comes from photosynthesis?

Do you remember that a seed contains stored nutrition and a plant embryo? One part of the embryo is called a **cotyledon** (cot-uh-lee-dun). The cotyledon becomes the first leaf of the plant. But before that, it gives food

Rice has a monocot seed.

Monocot plants grow in different ways than dicots do. Monocots have parallel veins in their leaves. You can tell that I came from a dicot plant because my leaf veins are branched instead!

Maple leaves have branched veins.

Lily of the valley leaves have parallel veins.

to the new plant until photosynthesis starts. Where does the cotyledon get the food it gives away? It absorbs it from the stored nutrition in the seed.

Scientists divide plants into two big groups according to the number of cotyledons in their seeds. Plants with one cotyledon are in the **monocot** group. Plants with two cotyledons are in the dicot group.

How a Seed Germinates!

1. Water and oxygen enter the seed once the seed coat allows it.

2. The seed swells and breaks the seed coat.

3. Water brings the waiting proteins and processes back to life.

4. Repairs are made to

DNA that may have been damaged because of age.

5. The seed makes new proteins and processes.

6. The baby roots grow downward out of the seed.

The hormone auxin tells roots to grow downward and stems to grow upward.

Bean Seed Germination

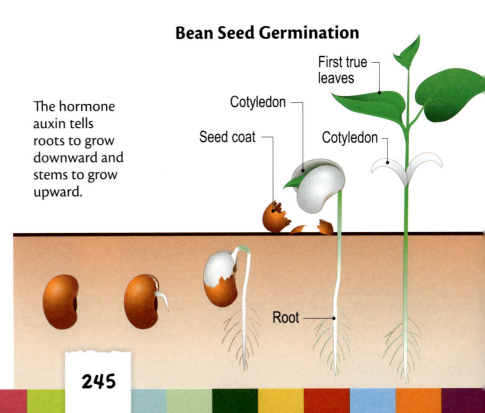

First true leaves

Cotyledon

Seed coat

Cotyledon

Root

7. The baby stem and leaves grow upward out of the seed.

8. The stem and leaves grow into the light and begin photosynthesis.

9. When the new plant can make all its own food, we say it's *established*.

Interesting Facts about Seed Germination:

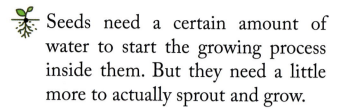 Some seeds leave their cotyledon underground as they sprout. Others bring it along and drop it later.

Along with water, the seed needs oxygen in order to sprout. Seeds can "drown" if the ground is flooded.

Dry seeds can be kept for hundreds of years and still grow when given water.

Seeds need a certain amount of water to start the growing process inside them. But they need a little more to actually sprout and grow.

If a seed has only enough water to start processes but not enough to sprout, the processes will use up all the stored nutrition, and the seed will never be able to sprout.

This Judean date palm was grown from a seed found in the country where Jesus was born. The seed was from a tree growing at the time Jesus was born about 2,000 years ago!

Pea sprouts leave their cotyledons underground.

Bean sprouts bring their cotyledons up with them to help with photosynthesis.

Prayer

Dear Jesus, we are amazed at the new life You bring to the earth through germinating seeds. We thank You for being the resurrection and the life for us to believe in, and for giving us life that goes on forever! Amen.

Time to do Activity 80 in the Activity Book!

Hollyhock seeds need light to germinate, so don't bury my seeds!

Today is Green Thumb Day! Time to do Activity 81 in the Activity Book.

CHAPTER 28
Special Plants in Special Places:
Aquatic Freshwater Biome

There are two types of aquatic biomes: **freshwater** and saltwater. It's easy to see the entire saltwater biome (oceans) on a world map. But we cannot see the entire freshwater biome. A world map can only show a few very large lakes and rivers, but it can't show the millions of other bodies of **fresh water** God has sprinkled around the world. Let's learn about the freshwater biome!

God has provided fresh water almost

A desert well in India

Definitions

We say "**aquatic**" when something is in or on the water.

Freshwater is a body of non-salty water like a lake or river.

Fresh water is water that isn't salty and is good for drinking if it's clean.

Natural spring

everywhere for thirsty people, animals, and plants. In areas with no water on the earth's surface, underground water can often be brought up from wells that people dig. And in some places, water naturally comes up from underground springs.

[God said,] "I will open rivers in desolate heights,
And fountains in the midst of the valleys;
I will make the wilderness a pool of water,
And the dry land springs of water.
I will plant in the wilderness the cedar and the acacia tree,
The myrtle and the oil tree;
I will set in the desert the cypress tree and the pine
And the box tree together,
That they may see and know,
That the hand of the LORD has done this.
And the Holy One of Israel has created it."
(Isaiah 41:18-20)

About the Freshwater Biome

Do you know how God brings fresh water onto the land? He brings it from clouds in the sky when it rains or snows. Do you know how water gets up into the sky? It evaporates from oceans, lakes, and rivers, and it also evaporates from plants during transpiration. Do you know how water got into oceans, lakes, rivers, and plants? It comes from

Can you find transpiration in this drawing of the water cycle? Find evaporation. Do you see where water has soaked into the ground and is running toward the lake? Someone could dig a well to that underground water.

rain and snow, and we are back where we started! This going-around of water is called the **water cycle**.

Across the land, we see water in different forms. We call these **bodies of water**. It's often helpful to put things in groups as we learn about them. Let's look at the different groups we use to talk about bodies of fresh water.

1. **Still water** is a body of water that stays in one place, though there is usually a slight current even in still water. This keeps it fresh and mixes in oxygen, nutrition, and different temperatures. Still water includes lakes, ponds, and wetlands.

   - **Lakes** are large bodies of water surrounded by land. They are large enough that their water can be mixed by wind and waves. Lakes are too deep for water plants to grow in them, except near the shore.

   - **Ponds** are smaller than lakes and are shallow enough that water

Beavers have made a pond in this conifer forest by building a dam of sticks across a creek. Then they made a domed home in the middle of the "beaver pond" that formed.

plants can grow on the whole bottom of the pond. But they are deep enough that those plants don't poke above the surface except near the shore. Ponds are too small to have waves.

   - **Wetlands** are flat areas of soggy soil that are covered with water at least part of the year. This water is shallow enough that plants stick out of its surface. Wetlands include marshes, swamps, and bogs.

Lake

Marsh

- Marshes are wetlands with no trees. The plants in marshes are usually monocots, like cattails and grasses. Most marshes are freshwater, but if a marsh is located near a seashore, it could be a salt marsh. In a salt marsh, salty ocean water mixes with fresh water running off the land.
- Swamps are wetlands with trees and bushes growing in the water.
- Bogs only receive water from rain or snow. Bogs only lose water through evaporation. This means bog water is completely still or **stagnant**. Rotting plants, low oxygen, and poor nutrition in bogs makes them unfriendly to most life.

2. **Flowing water** is a body of natural moving water that keeps flowing in one direction. Flowing water starts somewhere high on land. Because of gravity, it flows downhill and empties into a lower body of water.

   Flowing water includes rivers, streams, and creeks (although sometimes "stream" means any flowing body of water, including rivers and creeks). Scientists don't have a perfect way to tell the difference between rivers, streams, and creeks. Usually, **rivers** are the

Bog

Swamp

I always say that a creek is something you can jump over, and a river is something you have to cross by boat or by swimming!

"Brook" and "creek" are two names for the same thing.

widest and deepest; **creeks** are the narrowest and shallowest; and **streams** are somewhere in between. One stream might have the name Crystal River because it is large compared to nearby streams. But somewhere else, a stream of the same size might be named Crystal Creek because it's smaller than nearby streams. Creeks usually empty into rivers, but rivers don't empty into creeks.

And [the angel] showed me a pure river of water of life, clear as crystal, proceeding from the throne of God and of the Lamb. (Revelation 22:1)

Time to do Activity 82 in the Activity Book!

Special Plants in the Freshwater Biome

Lakes have areas or **zones** where the temperatures are different. The zone near the shore is the **littoral** (lit-er-uhl) **zone**. This zone is the warmest because the water is shallow and can absorb more of the sun's heat. Because a pond is shallow and small, the whole thing is warm like the littoral zone of a lake.

Both ponds and the littoral zones of lakes have beautiful aquatic plants—some floating and

Water lily flowers and leaves float on the water's surface at the end of their long stems. Their roots are secure in the lake bottom below.

some rooted. God gave these plants special gifts to help them live in water.

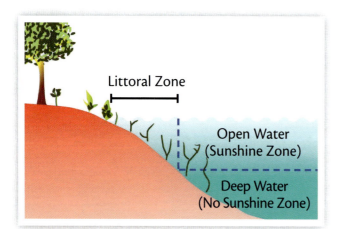

Littoral Zone

Open Water
(Sunshine Zone)

Deep Water
(No Sunshine Zone)

Eelgrass usually lives and grows entirely under water. God gave **submerged** plants the ability to use dissolved carbon dioxide in the water for photosynthesis.

Lake littoral zones and ponds also contain wonderful forms of life called **phytoplankton** (fie-toe-plank-tun). *Phytoplankton* is the word for all the kinds of microscopic, plant-like things in water. Scientists don't call these things plants, but they say they are like plants because they do photosynthesis. Phytoplankton include algae, diatoms, and bacteria. Phytoplankton are very important because they make oxygen

One type of single-celled freshwater algae as seen through a microscope. Notice its four green machines.

Aquatic plants like the yellow water lily often have tiny pipes inside them that pass air from their leaves at the surface to their roots in the soggy lake bottom.

The water hyacinth has balloon-like leaf stems to help its plant float while its roots hang down in the water.

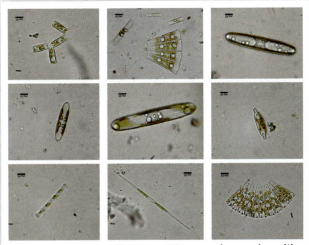
Living freshwater diatoms. Diatoms have glass-like cell walls.

Photosynthetic bacteria

and become microscopic food for microscopic creatures.

Away from the lake's shore, water continues to be warm near the surface. This area is called the **sunshine zone**. There are no large aquatic plants in a lake's sunshine zone because the water is too deep. But the number of phytoplankton is greater here than in the littoral zone.

The deeper you go in a lake, the colder and darker the water gets. After a certain depth, photosynthesis isn't possible because there is almost no sunshine. This is the aphotic zone, but we will just call it the **no-sunshine zone**. The phytoplankton living in the sunshine zone only last about six days. When they die, they fall down to the bottom of the lake and rot. But this is good! Just like on land, rotting plants

provide nutrition for other life. This nutrition circulates through the water and is used by aquatic plants and new phytoplankton.

Mountain marsh marigolds love the icy water that comes from melting snow! They bloom in early spring and are a favorite food of moose and elk.

The flowing water of creeks, streams, and rivers aren't always the same temperature. Water temperatures change as the water flows downhill. Rivers usually start as small creeks in higher, cooler areas. Creek water moves quickly as it tumbles down a steep valley. Aquatic plants can't grow in fast-moving water, but moss often covers the submerged rocks in these creeks and rivers.

River water becomes slower and warmer as it flows thorough gently-sloping valleys that aren't very steep. We can find some of the same plants at the edges of slow rivers that we find in the littoral zone of ponds and lakes.

Horsetail is a kind of rush. It likes poor soil in wet marshes.

Time to do Activity 83 in the Activity Book!

Red-winged blackbirds have floating nests that are loosely attached to cattails. When the water level changes, the nest rises and falls with it!

Special Animals in the Freshwater Biome

Ponds and lakeshores are home to many animals! Some live only in the water; others spend time in the water and on land too. Let's look at some of the animals you might see as you explore a pond or lakeshore.

Underwater, you might find a freshwater snail on a rock or a clam in the sand. Lift a rock, and you might

256

Family of Canada geese

find a dragonfly or mayfly larva that is able to spend its childhood underwater because God gave it gills. You may see frog eggs or tadpoles which will grow into adult frogs that spend time on land. You might even see a crayfish suddenly scoot backwards into a hiding place, leaving a cloud of dirt in the water. And of course, you may see fish!

It's fun to catch a crayfish. Beware of its pinching claws!

Other animals go in and out of ponds and lakes, but they stay near shore while on land. If an enemy comes near, they can jump into the water and swim safely

Duck

away. Some of the animals that are comfortable with both land and water are ducks, geese, beaver, muskrats, otters, frogs, turtles, and some snakes.

Two types of animals usually live in the sunshine zone of lakes: fish and **zooplankton** (zoh-uh-

River otters love to play in water!

257

Zooplankton

Turtles basking on a sunny log

plank-tun). Zooplankton include tiny microscopic and visible water creatures that can't make their own food with photosynthesis. Instead, they eat phytoplankton! Some of these creatures don't eat phytoplankton, but they capture them. They keep the phytoplankton inside their clear bodies and live off the food their prisoners

Trout need water with a lot of oxygen.

make using sunshine!

The no-sunshine zone, with its cold, dark waters, is not very friendly to life. But there is a very necessary form of life at the bottom of lakes. Anything that dies in a body of water sinks to the bottom. This is where microscopic creatures eat dead plants and animals and turn them into nutrition for other living things.

As creeks tumble down from high places, they pick up lots of oxygen as their water splashes through the air. Farther down, they join streams and rivers, and they slow down. They don't splash anymore, so the water has less oxygen in it.

Certain fish need more oxygen in water than others do. These fish like to stay in the top part of the creek and river. But downstream, there will be other fish who get

Catfish are content with murky, low-oxygen water.

along fine with less oxygen. There are many different plants and animals in the middle part of a river where it's wide. Then, near the end, the river levels out and runs even slower. The water is usually cloudy because of all the dirt picked up along the way. But there are a few fish who don't mind this kind of water at all! God has made special fish for each part of the creeks, streams, and rivers.

Prayer

Father, thank You so much for the fresh water You have put all over the world. Thank You for keeping lakes, ponds, wetlands, creeks, streams, and rivers full of water by making rain and snow. We praise You for the life You put in the secret underwater places. It's so fun to go fishing and see what comes up from under there! Amen.

Time to do Activity 84 in the Activity Book!

UNIT 8
Powerful Little Punches

God has put some powerful little punches in plants! Some keep you healthy. Others help you heal when you are sick.

And others give flavors to make food yummy!

Let's give God glory for the powerful little punches that give health, healing, and flavor to our food!

 ## Hymn to Sing

God, Who Made the Earth

God, who made the earth,
The air, the sky, the sea,
Who gave the light its birth,
Careth for me.

God, who made the grass,
The flower, the fruit, the tree,
The day and night to pass,
Careth for me.

God, who made the sun,
The moon, the stars, is He
Who, when life's clouds come on,
Careth for me.

God, who sent His Son
To die on Calvary,
He, if I lean on Him,
Will care for me.

You can listen to this hymn by searching for "God Who Made the Earth children's hymn" on the internet.

 ## Memory Verse

Therefore, whether you eat or drink, or whatever you do, do all to the glory of God.
(1 Corinthians 10:31)

CHAPTER 29
Nutrition Power

Do you remember that plants make all the carbohydrates that feed life on Earth? You may also remember that plants make some of the fat and protein we eat too. Carbohydrates, proteins, and fats are **nutrients** we need in large amounts.

But there are other nutrients we need in very small amounts. We call them **micronutrients**. Micronutrients are the vitamins and minerals we can't live without. They are necessary for all the chemical processes in our bodies, but our bodies can't make them. We need to eat them! That's why God gives plants the ability to give us the vitamins and minerals we must have! Let's look at some of the micronutrients. We'll discover a few of the foods containing them and why each micronutrient is necessary.

Definitions

A **nutrient** is something we eat that is necessary for growth and life. Nutritious foods contain many nutrients.

Whole grains are grains that haven't had their bran and germ removed.

Bran is the outer covering of a grain after the husk is removed.

Germ is a grain embryo.

God Gives Vitamins

- Vitamins are powerful little punches that are made by plants and animals.
- Each vitamin is a very complicated chemical.
- Vitamins are easily destroyed by heat or air.

Flora's List of Vitamins

- **Vitamin A** is found in green and yellow fruits and vegetables and in animal livers. Without vitamin A, our bodies could not use protein. Our eyes, skin, cells, bones, and teeth need vitamin A. Our bodies also need it to fight germs.

- **Vitamin B** is found in many foods, but especially in **whole grains**, nuts, vegetables, meat, and egg yolks. There are eight kinds of vitamin B that must

Which food do you think has more vitamins?

work together as a team. One of them, B12, can only be found in animal products, sea vegetables (like kelp), and brewer's yeast. The B vitamins are needed for our nerves, brain, eyes, skin, mouth, hair, blood, liver, and energy.

- **Vitamin C** is found mostly in plants, although raw meat has a small amount. Berries, green vegetables, and citrus fruits (like oranges and tangerines) have a lot of vitamin C. We need vitamin C for our bodies to grow, repair damage, and fight germs. Most animals make their own vitamin C in their bodies, but people and some animals cannot. Thankfully God gives it to us in our food!

- **Vitamin D** is a vitamin we make in our bodies when sunlight hits our skin! But God has also put vitamin D in foods since we sometimes can't get enough sunshine. Vitamin D is found in fatty fish, cheese, butter, eggs, vegetable oils, oatmeal, sweet potatoes, and dandelion greens. Children need vitamin D for their bones to grow properly. It is also needed for muscle strength, good blood pressure, fighting germs, making hormones, and helping our bodies to stop bleeding.

- **Vitamin E** is found in whole grains, nuts, seeds, beans, and dark green vegetables. We need vitamin E to protect our cells from damage—especially red blood cells and nerve cells. It repairs our bodies and keeps our blood circulating well.

- **Vitamin K** is found in grains, green vegetables, molasses, egg yolks, and liver. Without vitamin K, injuries would not stop bleeding, and bones would not form, grow, or repair. Vitamin K also has some jobs to do in our livers.

Time to do Activity 85 in the Activity Book!

God Gives Minerals

- Minerals are powerful little punches that are taken up by plants from the soil and given to us when we eat the plants. Minerals come from nonliving things like rocks.
- Minerals are very simple chemicals.
- Minerals are not easily destroyed.

Herby's List of a Few Minerals

- **Calcium** is found in milk products, green leafy vegetables, and seafood. We need calcium for our bones, teeth, muscles, nerve signals, and to stop bleeding.

- **Iron** is found in meat, eggs, dark green vegetables, raisins, and beans. Iron is used by our blood to make it possible for life-giving oxygen to be carried throughout our bodies.

- **Magnesium** is found in most foods. Many fruits, vegetables, nuts, and grains are high in magnesium. Magnesium is needed for 600 different processes including nerve signals and the proper beating of the heart. It helps muscles relax, including muscles in the blood vessels (which helps blood pressure). Magnesium is also needed by our bones.

- **Potassium** is found in bananas, dried fruits, vegetables, meat, and milk products. Potassium is in charge of nutrients passing into our cells. It balances the water in our bodies and is needed for nerve signals and heartbeats.

- **Sodium** is found in all foods and in our salt shakers. It works with potassium to balance water, and it makes adjustments to the blood.

- **Zinc** is found in pumpkin and sunflower seeds, beans, pecans, mushrooms, egg yolks, and meats. It's needed for making protein, fighting germs, and healing wounds. Zinc helps us taste and smell.

God has been so kind to give us little punches of power in these and many other foods. Since we need vitamins and minerals to stay alive, He has made it easy for us to get nutritious food!

**And He is before all things, and in Him all things hold together.
(Colossians 1:17, ESV)**

Prayer

Jesus, You were before all things. And now You hold all things together. You make everything in the world work perfectly together, and You make our bodies work with the good things we get from food. Thank You for giving us vitamins and minerals as powerful little punches so we can live! Amen.

Time to do Activity 86 in the Activity Book!

CHAPTER 30
Healing Power

God gave plants the job of feeding us. This verse shows that plants can also help our health!

When God created the world, everything was perfect. The first two people, Adam and Eve, never got sick. But after Adam and Eve disobeyed God, the world changed. Sickness and death came into the world. But God is kind to us! He gave us healing chemicals in plants. He also gave people brains to discover and invent **medicines** for healing. But best of all, He made a way for us to live forever after we die. He sent Jesus to die for us so we can be forgiven!

Definitions

A **medicine** is something we take into our bodies to heal us, prevent sickness, or help us feel better when we are sick or injured.

Phytochemicals (fie-toe-kem-ick-uhls) are chemicals a plant makes for its own protection. Phytochemicals give plants their colors and flavors. God also gives phytochemicals helpful jobs to do inside our bodies when we eat them.

Heal me, O LORD, and I shall be healed;
Save me, and I shall be saved,
For You are my praise.
(Jeremiah 17:14)

God can save and heal!

God Gives Phytochemicals

Plants make over 25,000 different chemicals so they can defend themselves against diseases, unfriendly weather, and hungry insects or animals. These **phytochemicals** are helpful to people too!

Phytochemicals are not necessary to keep us alive like vitamins and minerals are. But they can help us stay healthy and can help our bodies heal when we're sick. Phytochemicals are little punches of power that act in different ways. Many phytochemicals work to keep us healthy. Others attack enemies inside our cells. Some keep enemies away from our cells in the first place. If bad cells start growing, some phytochemicals can stop these bad cells from getting nutrition so they die. And there are others that can help our own enemy-fighting cells do a better job! Phytochemicals are powerful little helpers!

Phytochemicals are found in the most colorful parts of a plant, like its flowers, fruits, and dark green or colorful leaves. These complicated chemicals give color, taste, and smell to the plant. Although we don't know everything about these powerful helpers, we do know that people who eat a lot of different plants that are high in phytochemicals are healthier than people who don't.

Phytochemical is a big word, and each phytochemical has a different big word for its own name. To keep it easy, let's look at phytochemicals according to their colors instead of their names. You will see just a few of the many ways these powerful little punches can help your body!

Red and pink phytochemicals:

- protect DNA
- help heart and blood vessels
- help keep cells healthy

Yellow and orange phytochemicals:

- protect DNA
- help make vitamin A
- help bones, teeth, skin, and eyes
- help keep cells healthy

Green phytochemicals:

- protect DNA
- support eye health
- help keep cells healthy

Blue and purple phytochemicals:

- protect DNA

The colors of these fruits and vegetables are caused by their phytochemicals!

- may slow down aging
- help keep the heart healthy
- help keep cells healthy

White phytochemicals:

- fight germs
- help keep the heart and blood vessels healthy
- help keep cells healthy

This delicious recipe (right) would be good for your eyes! Corn and orange peppers each have two phytochemicals that are sent to your eyes to protect them from harmful blue light. Blue light is strong in bright sunlight and computer screens. The two phytochemicals can be destroyed by the blue light while they are protecting your vision. Eye doctors can see these yellow phytochemicals inside your eyes. If they don't see enough, they will ask you to eat the right foods to replace them.

Cocoa has many phytochemicals that may help keep our blood balanced and our hearts and cells healthy. Cocoa helps blood vessels in your brain and protects your skin from the sun. Chocolate is brown, not from its phytochemicals, but because it's made of cocoa beans that have been roasted. This dark chocolate is a healthier choice than milk chocolate because it has less sugar.

Time to do Activity 88 in the Activity Book!

God Gives Medicines

All around the world, God has placed plants that help with healing. Long before scientists could make medicine out of chemicals, people all around the world used these plants for medicine.

A piece of Chinese writing over

Chinese ginseng

5,000 years old reported on the healing benefit of Chinese ginseng (jin-sing) roots. The legend says that the emperor tried to kill a snake by beating it. A few days later, the snake came back, looking unhurt. He beat it again, thinking it would die. But it came back again. This time, the emperor followed the snake into a bush and watched as it ate part of the ginseng plant. Ginseng is still used today for injuries and many other problems. Scientists have found that its powerful healing punches come from phytochemicals.

Willow bark and leaves have been used for thousands of years to lower pain and fevers. Its **active ingredient** (the powerful little punch) is not a

Weeping willow tree

Henbane

phytochemical. It's the plant hormone **salicin**! Salicin can control growth and tell its plant when to start fighting a disease. This hormone can even travel from one plant to another through the stomata and tell the next plant it's time to fight a disease. Scientists have copied salicin to make aspirin for our pain and fevers. Farmers can spray this chemical on some crops to increase the amount of phytochemicals they make.

Henbane's powerful phytochemicals are used by the plant to keep it from being eaten. About 1,800 years ago, a doctor in India described over 700 different medicines from plants, including henbane. Henbane's phytochemicals are poisonous in large amounts, but in small amounts, they are just right to take away the poison of snakebites.

About 200 years ago, a scientist was able to do something new. He was able to extract from a plant the active ingredient that lowers pain. Then, about

100 years ago, scientists learned to copy the same chemical and make a pain medicine without using plants. This chemical is still used as a pain medicine around the world today.

Some of today's medicines are still made from plants, but most are manmade copies of plant chemicals. Manmade chemicals are easier and cheaper to make than medicines made directly from plants. Recently, doctors started using a few new medicines that scientists have invented using computers. But the scientists still start with the basic shape of the chemicals God created.

The bark of cinchona, a tropical South American tree, has been used for hundreds of years to fight a dangerous disease carried by some mosquitoes. The disease is called *malaria*. About 200 years ago, a scientist discovered a way to separate the healing phytochemical from the bark of this tree. Now, doctors use a manmade copy of the phytochemical to

Cinchona tree leaf and flowers. The bark of the tree is used to treat malaria.

things in the plant that will all work together to better heal this disease.

Sometimes a manmade medicine will cause problems as it does the job it's supposed to do. Amazingly, scientists can often go back to the plant the medicine started with and find something in it that prevents the problem! God does all things well!

Let's remember, the most important thing you can do during sickness is to pray:

And the prayer of faith will save the sick, and the Lord will raise him up. And if he has committed sins, he will be forgiven. (James 5:15)

treat malaria. Unfortunately, one form of malaria isn't responding well to the manmade medicine, so scientists are looking at the plant again. There might be more phytochemicals and other

Prayer

Thank You, Lord, that You gave us phytochemicals from plants. These are powerful little punches that help plants fight their diseases and also help us to heal. But we especially praise You for hearing our prayers! Thanks for the beautiful flavors and colors of the phytochemicals You give plants. Amen.

Time to do Activity 89 in the Activity Book!

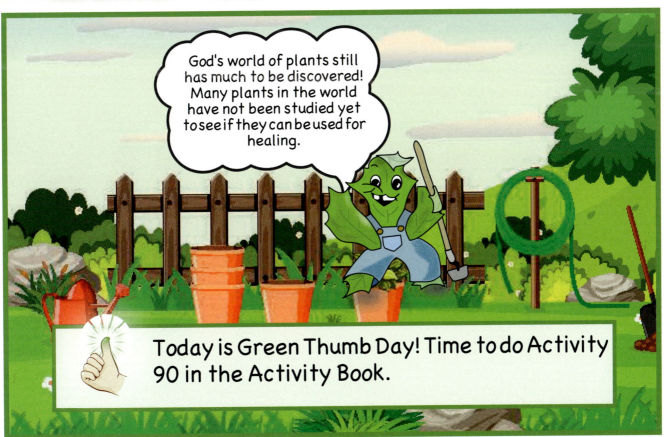

God's world of plants still has much to be discovered! Many plants in the world have not been studied yet to see if they can be used for healing.

Today is Green Thumb Day! Time to do Activity 90 in the Activity Book.

Basil is a delicious herb!

CHAPTER 31
Flavor Power

[Trust] in the living God, who gives us richly all things to enjoy. (1 Timothy 6:17)

Do you see in this verse that God is generous? He doesn't just give us boring things to supply our needs. He gives us what we need and also makes sure we enjoy those gifts.

God has given us the gift of food. And He has given us tongues to appreciate tastes like sweet, salty, and sour. But He has also added powerful little punches of flavor to make food even more enjoyable!

Do you remember that apples make fragrant chemicals under their skin when they are ripe? These fragrant chemicals are phytochemicals! And the lignins found in wood, which we've learned give flavor to barbecued meat, are also phytochemicals.

Sometimes we enjoy adding small amounts of powerful punches to our food to give it an interesting flavor. When we add spices to cake batter or herbs to spaghetti sauce, we are using them for their phytochemicals.

God Gives Herbs

In science, herbs are short, non-woody plants that die to the ground each year. But when we are talking about food, herbs are plants whose leaves we use for their flavorful phytochemicals. Because herbs are used in such small amounts, they are not called vegetables. Phytochemicals allow herbs to be used for flavor, fragrance, and health. You can plant some herbs and grow your own powerful little punches of flavor!

Most herbs can be grown inside or outside as long as they get a lot of sunshine. Their flavors are strongest before they flower, so use them or cut them before they bloom. Harvest herbs mid-morning, after the dew dries and before they wilt from heat. If you want to dry herbs, hanging them upside down in the shade is the way to help them keep the largest amount of phytochemicals. Here are a few common herbs you can grow.

Herby's List of Herbs You Can Grow

Basil

- likes rich soil that drains well.
- needs days with temperatures over 70° F (21°C) and nights over 50° F (10° C) before it will grow abundantly. (You may need to start seeds early in little pots indoors. When the weather is warm enough, the plants can be moved outside.)

Chives

- like well-drained soil that is rich but sandy.
- need the cool weather of spring or fall. They often go dormant in summer.
- will come back from the roots. They will also drop their seeds and make more plants for you!

Once basil is six inches (15cm) tall, cut off the top of the plant just above a pair of leaves. This will make two new branches and a bushier plant. Cut stems this way anytime you would like some tasty leaves.

In addition to the leaves, the pinkish-purple chive flowers are also edible.

Cilantro leaves and coriander seeds are an herb and a spice from the same plant!

Dill is fun to grow because it attracts a lot of pollinators. This dill-eating caterpillar will become a black swallowtail butterfly.

Cilantro

- is a fast-growing herb that likes any kind of soil.
- grows its leaves early, then flowers later in hot weather. Cilantro makes seeds (called *coriander*) which you can harvest as a spice when they turn brown.

Dill

- likes rich, well-drained soil.
- does not like to be transplanted, so don't start it early indoors.
- leaves, flowers, and seeds can be used.

Lavender

- likes poor, well-drained soil. If soil is too moist, lavender will have less flowers with less fragrance and may die.
- seeds can be planted indoors, 10 weeks early, by sprinkling them on top of soil. Lavender seeds need light to sprout. You can also buy plants from a garden center.
- may not bloom its first year.
- comes back healthier the next spring if new (not woody) growth is pruned to shape the plant.
- should live about ten years if the soil isn't too wet.
- may not bloom its first year.
- comes back healthier the next spring if new (not woody) growth is pruned to shape the plant.
- should live about ten years if the soil isn't too wet.

To dry lavender flowers, cut stems as the bottom flowers begin to open. Hang in a shady place.

Time to do Activity 91 in the Activity Book!

God Gives Spices

Spices are little punches of flavor power that come from the roots, flowers, fruits, seeds, or bark (but *not* the leaves) of plants. Spices are used for flavoring, dyes, fragrance, health, and to preserve food.

We know that one of the jobs of phytochemicals is to kill germs in their own plant. Spices contain extra-strong phytochemicals so they have extra germ-killing power. It's not surprising that God causes most spices to grow in tropical areas where hot temperatures and moist air make germs and mold grow quickly. Spice plants are able to quickly kill germs that try to infect them.

Germs and mold also make food rot. Before refrigerators and freezers were invented, one of the ways people **preserved** food (or made it last longer) was to mix in spices that kill germs.

Long ago, people that lived in the cool areas of the world didn't know what spices tasted like. But when they tasted

We usually need less of a spice to flavor food than we do of an herb because spices are so strong!

the spices brought to them by travelers and traders, they wanted more!

Explorers rushed to find new ways to get to the places where spices grow. And while they searched, they discovered people and lands they had never heard about! They brought back a lot of knowledge about the earth. They also brought back news that there was a world full of people that needed Jesus! Thanks to God's spices, missionaries were sent out to spread the gospel.

Let's look at some common spices and where they first grew before people spread them by planting them in other tropical places.

Flora's List of Spicy Places

Indonesia (a nation of islands in the Pacific Ocean)

- Some of the islands of Indonesia used to be called the Spice Islands.

- The Italian explorer Christopher

Nutmeg seed is grated to make spice.

Spice Islands, from antique Spanish atlas

Columbus hoped to sail across the ocean from Europe and travel around the world to reach the Spice Islands. But North and South America blocked the way!

- *Cloves* and *nutmeg* first grew only in Indonesia.

India

- *Cinnamon*, *turmeric*, and *black pepper* were first discovered in India and nearby tropical areas.

- Cinnamon is the bark of a tree, turmeric is an underground stem, and black pepper is a fruit.

Ginger can be used fresh or can be dried and ground.

Turmeric is an underground stem.

first grew in Central Asia, which is not a tropical area.

South America

- *Cocoa* (seeds), *paprika* and *cayenne* (fruits), and *vanilla* (seed pods of orchids) started in South and Central America.

Black pepper is a fruit that's picked green but turns black as it dries.

Cayenne peppers

Asia

- *Ginger* (an underground stem) and *licorice* (a root) are spices that first came from southern Asia.

- *Garlic* (an underground bulb)

North America

- North America is not in the tropical biome, but it's the first home of *sassafras*. The root bark of the sassafras tree is used to make root beer.

The sassafras tree has leaves with different shapes.

Prayer

Father, we praise You for the many wonderful flavors You give us in the powerful punches of herbs and spices. We can get so many different tastes by adding a different spice! We can take ground beef and tomatoes and make Italian pasta sauce, Hungarian goulash, or Mexican chili just by adding different herbs and spices! Thank You! Amen.

Time to do Activity 92 in the Activity Book!

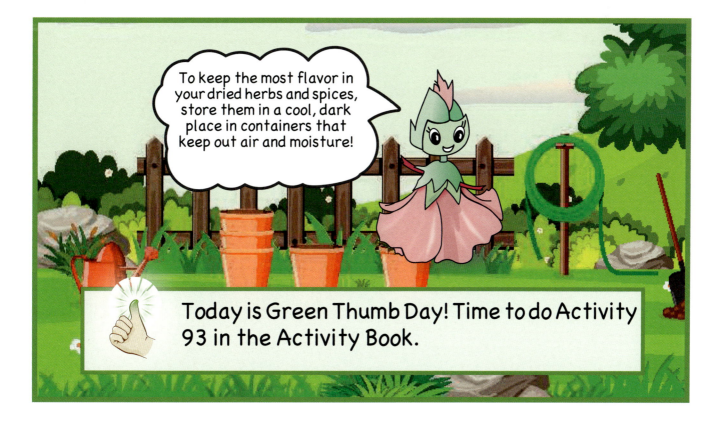

285

CHAPTER 32

Special Plants in Special Places:
Oceans

The sea is His, for He made it. (Psalm 95:5)

About the Saltwater Biome

God's sea (or ocean) is a very large body of water. It's so large that it covers over twice as much of the earth's surface as the land does. The earth's ocean is its **saltwater** biome.

Why is ocean water salty? The answer is found in Ecclesiastes 1:7:

All the rivers run into the sea,
Yet the sea is not full;
To the place from which the rivers come,
There they return again.

We have already learned about two things that will help us know why seawater is salty: the water cycle and saltwater. This verse from Ecclesiastes is talking about the water cycle—rain falls on the land and runs into bodies of water. As water flows over rocks, dirt,

Sea falls

Rainwater is fresh, not salty. God knows it's just what we plants and you humans need!

and plants, it picks up tiny amounts of salt. It's not enough salt for us to taste, but this salt travels in the water until it reaches the ocean.

We also learned that, if you could separate saltwater into smaller and smaller pieces, you would end up with water molecules and salt molecules. The water cycle separates water and salt for us! In the water cycle, when water evaporates from the ocean's saltwater, it leaves the salt behind. This has made the ocean become saltier over the years. The water molecules leaving the ocean travel up to become clouds that will rain on the land again.

 Time to do Activity 94 in the Activity Book!

Special Plants in the Saltwater Biome

What do you think of when you think of the big, wide ocean? Do you think of plants? Or do you think of huge whales, large fish, and giant octopuses swimming around? We have a hard time picturing plants in the ocean other than a little seaweed near the shore. The ocean seems more like a place for mysterious creatures than somewhere for plants to grow.

But these creatures have to eat something. Just as on land, there must be something (like food-making plants) to start the ocean food chain. God has supplied the perfect answer by filling the ocean with tiny plantlike things that are busy making food from sunlight. Tiny creatures eat the plantlike things, not knowing something bigger is going to swim up and eat them! None of the

Phytoplankton use sunlight, water, and air to make over half the world's oxygen!

creatures in the ocean would be alive if it weren't for these important, plantlike things starting it all. We have met their cousins before in the freshwater biome. These **photosynthetic** living things are our friends: the microscopic phytoplankton!

Phytoplankton are helpful friends because they supply the food that eventually gives us fish we can eat. But they are also helpful because God has given them the job of making oxygen. Just like land plants, phytoplankton use sunlight, water, and air to make oxygen. And they make a lot of it! If you weighed all the phytoplankton in the ocean, they would weigh as much as the plants of the tropical rainforest biome. But these phytoplankton make over half of the oxygen on Earth!

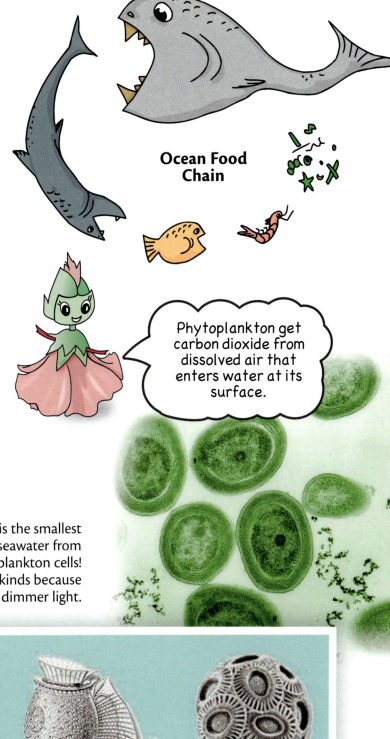

Ocean Food Chain

Phytoplankton get carbon dioxide from dissolved air that enters water at its surface.

This kind of phytoplankton (right) is the smallest photosynthetic thing on Earth. One drop of seawater from the surface could contain 7,000 of these phytoplankton cells! These phytoplankton can live deeper than other kinds because they are able to do photosynthesis using dimmer light.

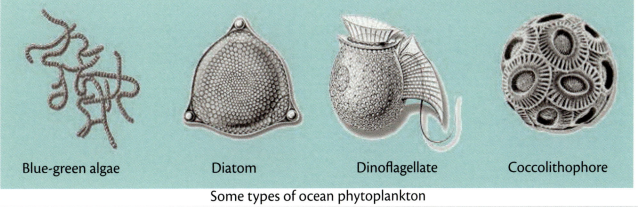

Blue-green algae   Diatom   Dinoflagellate   Coccolithophore

Some types of ocean phytoplankton

Ocean Zones

OARWEED

CORALS

ORCA

PLANKTON

JELLYFISH

SUNSHINE ZONE

SNAIL

WHALE SHARK

ORANGE ROUGHY

SPERM WHALE

TWILIGHT

OCTOPUS

SEAFLOOR

LEATHERBACK TURTLE

GULPER EEL

SPONGE

SEA CUCUMBER

ANGLERFISH

NO LIGHT

Almost all ocean life, including phytoplankton, live in the sunshine zone, especially in shallow areas near shores and coral reefs.

Besides phytoplankton, the ocean contains other interesting photosynthetic things:

Seaweed is not really a plant. It's an algae. Seaweed doesn't have roots or other plant structures.

Kelp is a type of seaweed. It's an important kind of ocean algae. Kelp lives in cool, shallow waters and provides a good home for thousands of different ocean animas. Tall "forests" of kelp provide food, shelter, and a place for many young sea creatures to grow up.

Kelp has structures called **holdfasts** to attach itself to rocks. It also has gas bladders (structures filled with air) to help it float upward as it grows. Kelp can grow quickly—about 10 inches (25 cm) in a day!

Seagrass is the only flowering plant growing in the ocean! It's a true plant,

Kelp is yellowish brown because of a certain phytochemical that helps it gather light.

Seaweed holdfasts

air. But recently, they found one kind of seagrass that makes nutritious, slimy blobs of pollen to attract creatures the same way nectar attracts pollinators on land! The seagrass-pollinating creatures are little crabs and ocean worms.

Like kelp, seagrass "meadows" are important homes for ocean animals. These meadows protect many young fish that will leave to become adult coral reef fish or nutritious fish for us to eat. Seagrass slows down rough waves, cleans the water, and gives food and shelter. Its spreading roots help keep the ocean floor from eroding. Sea grass grows in shallow waters near the shore and can grow to depths of 75 yards (70 m) if there is enough light from above.

with roots, leaves, flowers, and seeds. Seagrass is found in warmer parts of the world where kelp doesn't grow. For a long time, scientists thought that seagrass only released pollen in the water the way grass on land releases it into the

Batfish in seagrass meadow

Time to do Activity 95 in the Activity Book!

Special Animals in the Saltwater Biome

So God created great sea creatures and every living thing that moves, with which the waters abounded, according to their kind, . . . And God saw that it was good.
(Genesis 1:21)

Let's look at some of God's many amazing ocean creatures!

Seahorses in the Seagrass

A seahorse is more than just cute. God has made it a collection of surprises!

Father seahorse with a bellyful of babies

Flora's Collection of Seahorse Surprises

- For a long time, scientists couldn't decide what kind of animal a seahorse is. Even though it has no scales, they finally decided to call it a fish because it has gills and a swim bladder.

- Seahorses have heads that look like a horse's head. They use their long snouts to vacuum up tasty little creatures.

- A seahorse must constantly eat snacks because it has no stomach to store a big meal.

- Seahorses swim with wing-like fins on their backs. Even though these fins flap 30-70 times in a second, the seahorse is very slow. In fact, the dwarf seahorse is the slowest fish in the world. It can only move 5 feet (1.5 m) an hour.

- Seahorses have monkey-like tails that curl to hold onto things.

- Instead of an inner skeleton, seahorses have bony plates just under their skin.

- Seahorse mothers lay from 100 to 1,000 eggs inside the kangaroo-like pouch on the father's belly. After two to four weeks, the father squirts out a swarm of newly-hatched babies.

GOD MADE PLANTS

Spines in the Kelp

Kelp forests are thick and tall, but they have an enemy. Sea urchins love to eat algae! Sea urchins are found in the ocean all around the world—from warm tropical waters to the cold waters of the poles. They live at all depths, from rocky shores down to the bottom of the deepest trenches. But if too many of them live in a kelp forest, they can destroy it.

Thankfully, sea urchins are very tasty to sea otters. God made sea otters with their own built-in tables. While floating on their

> *Urchin* is an old word for *hedgehog*. Why do you think urchins have this name?

backs, they use their tummies as tables, setting their food on them as they prepare and eat it! To eat a sea urchin, the sea otter quickly spins it around in its paws, breaking off the spines. Then it uses its strong teeth or a rock to crack open the shell and lick out the insides. By eating urchins, sea otters help keep kelp forests alive to provide homes for many sea creatures. Otters often wrap a piece of kelp around themselves and their babies to keep them from floating away.

Sea otter with sea urchin

294

Did you know that sea otters have spines too? You can't see their spines without a powerful microscope. Their spines are on their hairs! Each hair is covered with tiny spikes that make the hairs catch on each other and give the otter messy hair.[4]

For the sea otter, messy hair is good! Sea otters live in cold water and don't have fatty blubber under their skin to keep them warm. Instead, they have the thickest hair on Earth—a million hairs in a square inch (6.5 cm²). But it's not the hair that keeps them warm. It's the air they can trap next to their skin with their hair. Since the hair is waterproof, and since the hairs can grab each other with their spines and make tangles, enough air is trapped to keep sea otters warm, even in cold, icy water.

Little and Big Filter Feeders

A **filter feeder** is an animal that feeds by passing water over a special filtering structure to strain out pieces of food.

Krill are little filter feeders that eat phytoplankton by using their front legs as filters. A krill's front legs have comb-like structures that, with its legs, form a "fishing net" as they swim through the water. When the net is full, the krill squeezes out the water. Then it can comb the phytoplankton into its mouth.

Krill are important little creatures:

Other predators of sea urchins are wolf eels, California sheepshead fish, and this crab.

Krill are only about half an inch (1 cm) long, but God has great purposes for them!

- Krill can gather and eat phytoplankton, something no other animal its size can do. They are at the beginning of the ocean's food chain.

- There are a lot of krill in the ocean! The total weight of krill in the ocean is more than the total weight of any other animal on Earth. Half of the ocean's krill becomes food for other animals!

- One, two, or three times a day, krill travel up and down in the water. They feed on phytoplankton at the surface. Then, when their tummies are full and they don't feel like swimming quickly, they sink. While they are in the depths, they let out their waste. Because they aren't so heavy anymore, they feel like swimming to the surface again. The up-and-down motion of so many creatures helps mix ocean water. By traveling up and down, krill become food for creatures that live at different levels or hunt at different times. And for nighttime hunters, they glow in the dark! Krill are helpful because their waste provides nutrition to deep plankton like manure does for plants.

- Krill stay in swarms. If you saw a swarm of krill, and had a bathtub for a scoop, you could take a scoopful of the swarm and have 2,000-9,000 krill critters in that tub!

Blue whales are *big* filter feeders. What do they eat? They eat krill, the *little* filter feeders. A blue whale will swim up to a swarm of krill, open its mouth, take in a mouthful of krill and water, and squeeze the water out, leaving the krill behind. The krill stay behind because they can't get past the whale's filter or **baleen**. Baleen are plates arranged along the gums like teeth on a comb. Each one is covered with baleen hairs to help catch krill. From phytoplankton to krill to blue whale is a very short food chain!

- Blue whales are the largest animals on Earth. They weigh up to 200 tons (180,000 kg). It's amazing that they can eat enough tiny krill to grow so

Krill

Blue whale showing baleen

mother's milk for over a year. And a baby whale needs almost 150 gallons (550 L) of milk every day! Mother whales need to eat a lot of krill to make milk for their babies!

large! A blue whale can eat more than 700 pounds (3,500kg) of krill a day.

- Whales are **mammals**. This means that a mother whale's body makes milk as food for her baby. Blue whale babies don't eat anything but their

Gulls and sea lions (like this bull, cow, and calf) are at home on the ocean's shore.

Prayer

Lord, thank You for the ocean biome with so many kinds of mysterious creatures. Thank You for supplying them all with food through phytoplankton. And thank You for giving us the oxygen we need through phytoplankton! You are so wise to make things work well in the ocean and in the whole world. Amen.

Time to do Activity 96 in the Activity Book!

297

You Can Plant a Garden!

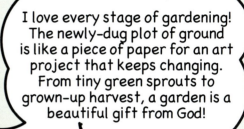

Now comes my favorite part! I think it's great that the first job God told people to do was to take care of His Garden of Eden. Now you can take care of a garden too. Let's go outside and get to work!

I love every stage of gardening! The newly-dug plot of ground is like a piece of paper for an art project that keeps changing. From tiny green sprouts to grown-up harvest, a garden is a beautiful gift from God!

About Your Garden

These next chapters will guide you in growing a small garden of your own! You will learn how to grow some typical, easy-to-grow foods. As you have learned, many different plants grow all around the world. You may like to try growing different plants than the ones covered in this unit if you know they do well in your area. This book teaches about gardening in places that have cold winters and warm summers. If you live somewhere that is always warm or hot, your garden may do better in the cooler season.

To grow a typical vegetable or flower garden, you will need a plot of ground that gets at least six hours of sunlight a day. If that's not possible, maybe you can find a sunny balcony or porch that will hold a few large pots. You can even grow some food plants indoors with a sunny window and special grow lights. Small varieties of plants are often available for gardening in pots. Even if you have no place for a garden right now, you can learn about gardening by reading this unit.

A garden is a worthwhile project with a lot of fun jobs spread over several months. There might be days when you spend an hour or two gardening, and there may be other days when you just glance at your garden to see its progress. Sometimes, you may just want to sit and admire it as part of God's creation.

Unit 9 will guide you in making a small garden. You will need help from an adult. A garden takes many months to grow from bare soil to harvest. Now and then, you may need to interrupt your progress through Units 1-8 of this book in order to work on Unit 9's garden. Or you may want to keep a steady schedule through the book and work on your garden after school and on weekends.

Here is a schedule of when you might expect to do certain gardening jobs as explained in the next chapters:

At beginning of this course:

- Make compost bins

Fall:

- Mark out an 8 x 8 foot (2.5 x 2.5 m) garden plot
- Clear away weeds and rocks
- Dig up soil
- Cover with a winter blanket of plant material

Winter:

- Seek advice
- Figure out your planting zone
- Plan your garden
- Order seeds
- Test soil (optional)

Late winter to early spring:

- Start seeds indoors

Early spring to early summer:

- Prepare soil
- Plant seeds outdoors
- Transplant small plants outdoors

Spring to fall:

- Water
- Weed

Summer to fall:

- Harvest
- Eat

CHAPTER 33

Prepare and Plan a Garden

Sow for yourselves righteousness;
Reap in mercy;
Break up your fallow ground,
For it is time to seek the LORD,
Till He comes and rains righteousness
on you. (Hosea 10:12)

I like the action in this verse! When it's time to work in our gardens, let's remember it's always time to seek the Lord.

Make Your Own Compost

Do you remember that God makes dead plants help the soil? Dead plants give nutrition to future plants. This nutrition is added to the soil as small creatures (visible and microscopic) break down dead plants by eating them. Dead plants feed the creatures, and the creatures help new plants by improving the soil! Without these creatures, dead plants would pile up, and soil would become poor.

We can help our soil by making **compost** all year long using kitchen

Definitions

Compost is made of things that were once living but now are rotting with the help of small, visible creatures and microscopic creatures. Compost is used as a plant fertilizer. When compost has been completely broken down by microscopic organisms, it's called *humus*.

and yard waste. You may want to start making compost right away so that you can use it in your garden! Here's how:

1. Make a compost bin.

When we make compost, we are speeding up the rotting process God designed to take care of dead things. One way to make compost is to build or buy a compost bin to hold the pile together. A bin keeps the compost from drying out too quickly or blowing away. The bin also helps hold in the heat made by the microscopic creatures as they break down the compost. Heat speeds up the process of composting!

Compost creatures need oxygen, so your bin should have holes or slits in the sides. Your bin can be round or square. You can make a bin out of wooden pallets, a strip of snow fence, or cinder blocks. Or you can make one out of a wire mesh (like chicken wire), but this may let in too much air, making your pile dry out too easily. Composting happens more quickly in large piles, so make your bin about 3-4 feet (1 m) across.

Be sure to set up your bin on the ground rather than on concrete. This allows worms to come and go into the compost when its moisture and temperature are best.

Anchor your bin to the ground so that you won't knock it over when you work with your pile.

Compost bin made of recycled wood pallets

Compost bin made from wire and plastic mesh

If you live in a place that gets a lot of rain, you may want to make a lid or roof over your bin to keep your compost creatures from drowning.

2. Build your compost pile.

To allow for more air at the bottom of your pile, make a loose first layer of stiff material like thin branches or cornstalks.

Next, make thin layers of smaller material. Compost is made quickly when you use a mixture of live green material and dry brown material. If you can, use twice as much brown as green. To avoid attracting pests like rats and flies, don't add oils or animal products. Egg shells are fine to use.

Worms in compost

If you start with the right mixture of green and brown in your compost, and if you stir it with a pitchfork two weeks later, you might have finished compost in two more weeks! That's only a month for a watermelon rind to become fertilizer!

Green Material

- Lawn clippings
- Kitchen produce scraps
- Annual weeds without seeds (Don't use perennial weed roots. They will sprout again.)

Brown Material

- Raked leaves
- Dried, disease-free yard and garden plants
- Small amounts of hay, straw, sawdust, wood chips, and coffee grounds

Layer these green and brown materials. Sprinkle a little dirt over every few layers to add microscopic compost creatures.

As different creatures work on your compost pile, it will heat up and then cool down. The heat helps kill weed seeds. After your compost cools, the worms find your pile and start their job of breaking down the compost.

If you want to help the process, you can add more worms. To attract worms, lay a piece of wet cardboard over a patch of damp soil. Check underneath it for worms in a day or so. Then add the worms to your compost bin.

Keep adding green and brown plant material over the next couple of months.

3. Water your compost.

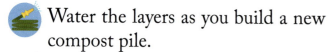 Water the layers as you build a new compost pile.

Compost should feel like a damp, squeezed-out sponge. Add water whenever it's too dry. Water keeps the compost creatures alive. Too much water may drown them. If you have a lot of roly-poly (sow or pill) bugs in your pile, it's probably too wet. If you have an ant colony in your pile, it's too dry.

4. Mix your compost.

Every few weeks, stir your pile by lifting and turning the compost with a pitchfork. Add water as needed.

If you have two compost bins, you can start a compost pile in one of them. When that pile is partly finished, move it to the empty bin. This will mix it, and you can start a new compost pile in the first bin. A third bin could be used as a place to store finished compost.

Unfinished compost

Finished compost

5. Use your compost.

Your compost is ready to use when it's dark brown and you no longer see pieces of material you can recognize. If you mix unfinished compost into your soil, it will take energy away from your plants and give it to the rotting process. Be sure your compost is ready before you put it on your garden.

Spread compost up to two inches deep when you use it on your garden.

Prepare Your Garden Plot

Fall is a good time to prepare your garden for next spring! If you can't do it in fall, late winter or early spring is fine. An 8 x 8 foot (2.5 x 2.5 m) square is a good size for your first garden.

1. Find a spot for your garden!

Vegetable gardens need six or more hours of sunshine a day. Be sure to check on the area you're thinking of using during the day to make sure it gets enough sun.

Choose an area safe from the wind, if possible. If not, you can build a windscreen near the side of your garden from which the wind usually comes. Windscreens should not be solid because wind will climb up and over solid things, giving plants a blast where it comes down. A snow fence or recycled wood pallets may be good, inexpensive windscreens. Be sure your windscreen is not close enough to shade your garden. A four-foot tall (1.2 m) windscreen can calm the wind for 80 feet (24 m)!

Make sure the plot you've chosen is safe to dig in. Have an adult help you search on the internet to find a free service that will mark underground pipes and wires. You want to avoid these!

2. Dig up the dirt!

Fall is a good time to prepare your soil for spring. If you wait to dig until spring, you will need to wait until the soil dries from the winter moisture. And after digging, you should let the soil settle a week or two before planting. That's why digging in the fall may allow you to plant sooner in the spring! It's good to cover soil with compost, leaves or other plant material over the winter. This keeps soil nutrition from washing away and keeps soil life from drying up.

Before you dig, let's figure out what kind of soil your garden plot has! Squeeze a handful of damp soil into a clump. Toss it up and down in your hand like a ball.

If it falls apart after two or three tosses, you have **loam**, a good mixture of sand and clay.

If it falls apart in the first toss and feels gritty, you have **sandy** soil. Sandy soil doesn't hold onto water and nutrients, but you can add compost to help improve the soil.

If the clump stays together no matter how long you toss it, you have a **clay** soil. Clay holds onto water and nutrition, but it clumps together when wet. You will need to avoid digging when clay soil is wet to keep it from getting packed. You can make paths or use steppingstones so you won't pack your garden's soil by stepping on it. Adding compost also helps clay soil.

Get a shovel or spade. It's time to dig! We dig up ground to loosen the soil so roots can grow deeper. If your garden plot is on ground that hasn't already been gardened, you should dig deeply the first year. After that, you may only need to dig deeply every few years. Deep digging is more important for heavy clay soils than for loam or sand.

Soil creatures need air. That's

why they live in the top few inches of soil (the **topsoil**). Before you dig deeply, remove 8-12 inches (20-23 cm) of topsoil from your small plot and put it to the side on a tarp or in a pile. We do this so we won't bury helpful creatures where they can't breathe. Try not to chop at the soil too much. You might kill worms and other soil life. It's okay to leave the soil in lumps. Lumps should settle if you wait long enough to plant.

Now loosen dirt under the topsoil (the **subsoil**) by turning it over in its place with your shovel. Replace the topsoil on top of the loosened subsoil. Top with a winter blanket of compost or other plant materials.

If you are deep digging in the spring, mix compost into the topsoil as you put it back into your plot.

If you are not deep-digging, you can just loosen the topsoil with a shovel or garden fork each year. Add compost on top before you work with the soil, and it should mix in. Try not to disturb soil very much. The main purpose of digging in the soil is to add air for its creatures.

Plan Your Garden

Winter is a good time to plan your garden. It's fun to think about green, growing things when the soil is cold and resting. Here are some things you can do to plan!

Visit a plant nursery. The nursery workers are usually gardeners too! They enjoy their jobs and like giving advice. They can tell you what plants grow well in your area. They can prepare you for common problems

with your area's soil or climate.

Find out the number of the **hardiness zone** where you live. A hardiness zone is an area on a map where certain kinds of plants can grow because the climate is right. This will help you know when to plant certain seeds and plants.

Winter is a good time to send some of your soil to be tested. You can find out if your soil is low in certain things so they can be added before planting in the spring. These tests do have a cost, so you could wait to see if your garden has problems the first year and test next winter.

Draw a plan for your garden. Plan to grow tall things like corn where they won't shade the rest of the plot. If you live north of the equator, tall plants should be planted on the north side of your plot because the sun shines slightly from the south all day long. If you live south of the equator, your

Sample garden plans. With an 8 x 8 foot garden, you can have four, 3 x 3 foot (one square meter) planting squares. You could have two paths crossing in the middle and a planting square in each corner.

tall plants should be on the south side of your plot.

 Buy or order seeds. If you are interested in saving seeds from the plants you grow, be sure to buy **heirloom** or **open pollinated** seeds. This means the seeds you save will grow into plants like the plants they came from.

 Save your plan to look at next year so you can remember where you planted your crops. Plant next year's seeds in different spots within your garden. Soil gets low on nutrients when the same kind of plant lives in it every year.

To help you decide what to plant in your garden, you can get some ideas from the next few chapters!

Prayer

It's exciting, Lord, to plan and prepare a garden! We want to see how Your creation works to make food. Help us be diligent in the work You have given us to do. Amen.

Definitions

Open pollinated or **heirloom** seeds are seeds that have been naturally pollinated. The plants growing from them will be like the parent plants.

Hybrid seeds are made from a plant that people have pollinated with pollen from another plant of the same kind. They do this in order to get a certain trait. They might want to grow a small purple pepper plant, so they pollinate a small pepper plant with pollen from a purple pepper plant. But if you plant the seeds from the small purple pepper plant, they may not make the same kinds of peppers.

Genetically modified seeds were made by taking DNA pieces from a different kind of plant or creature to try to grow a better plant. For example, copies of DNA have been taken from a kind of bacteria and put into corn DNA to create something that can kill worms that eat the kernels. And there are some crops whose DNA has been changed so they don't die when they are sprayed with a strong plant killer. Only the weeds die. This helps farmers, but it means the crops we eat have strong plant killer on them.

CHAPTER 34

Sow and Grow Early Produce

God has wisely blessed us with a variety of vegetables and fruits that are ready to be harvested at different times of the year. This way, we can eat fresh, healthy food from our garden in spring, summer, and fall. And He also made some foods that store well so that we can stay healthy and fed all winter too!

The eyes of all look expectantly to You, And You give them their food in due season. (Psalm 145:15)

In the next few chapters, you will learn how to grow some common garden plants. In this chapter, we will start with vegetables you can plant and eat early in the growing season. The following vegetables grow best in the cool weather of spring. Their seeds can be planted directly into your outdoor garden.

Garlic, Lettuce, Radishes, and Peas

Garlic is an early food you can plant. It can even be planted in the fall so its roots can begin to grow before winter. Planting garlic in the fall instead of spring will produce larger bulbs when you harvest next summer. Plant at least one month before the ground might freeze.

1. Buy a bulb of garlic from a grocery store.

2. Separate the **cloves** (the sections making up the bulb). Each clove will grow into a new garlic bulb!

3. Plant garlic cloves four inches (10 cm) deep and four inches apart. Plant them with their pointed ends up. Bigger cloves grow into bigger bulbs, so don't plant little ones.

Garlic

4. Grass-like leaves will come up in early spring. You can cut a few leaves to use for a mild garlic flavor in food. But don't cut very many leaves, or your harvested bulbs will be small.

5. When a stiff stem shoots up in the summer, watch for a curly seed head starting to form on the end. Cut off the stem (**scape**) so that more energy goes to the bulb forming underground. But don't throw away the scape! You can stir-fry the tops of garlic scapes with butter and salt for a mild garlic treat to put on a salad or burger.

6. When the garlic leaves turn yellow, harvest the bulbs by carefully digging them up.

7. Lay out the garlic bulbs in a dry, shady place for four days to toughen the skins. After they are dry, trim the remaining roots close to the bulb.

8. Store your garlic in a basket or mesh bag in a cool, dark area.

Lettuce will come up very early in the spring. Remember, lettuce seeds can tell when spring has come by the amount of light they sense. Plant them early, and they will sprout at the time God has set for them!

1. Sprinkle seeds about two inches apart (5 cm) on top of fluffy, damp soil in early spring.

2. Sprinkle 1/8 inch (3 mm) of soil over the seeds. This will still allow the seeds to sense light.

3. Keep the soil damp until seeds sprout.

4. Plant more seeds every three weeks so you can eat salads for a long time. Lettuce likes to be in shade part of the day, so you can plant it between other plants. It won't harm them because its roots are shallow, and you will cut it soon.

5. When you pick lettuce, cut off the head a little above the ground. New leaves will grow out of the stem's stump for you to gather later. Pick lettuce early in the morning and refrigerate it so it will be crisp for your lunchtime salad.

6. When the weather gets hot and dry, lettuce will **bolt** (get tall and taste bitter). It will then grow yellow flowers and fluffy seed

Lettuce

Lettuce seeds

puffs. You can collect the seeds in a paper bag. When you are sure they are dry, store them in an airtight container and plant them next spring!

Radishes are one of the fastest sprouting seeds you can grow. You can see their first little leaves about a week after planting.

1. Radishes come in different varieties that can be planted at different

Radishes

times. Check the seed package for your planting time.

2. Plant radish seeds 1/2 inch (2 cm) deep and 3 inches (8 cm) apart. You can plant them in the same row or bed as other vegetables since they will grow fast and be harvested early.

3. Harvest radishes when they are less than one inch (2.5 cm) wide. If they grow bigger than that, they will get woody and taste very spicy.

Pea seeds need to have cold temperatures to germinate. To make sure this happens, you should plant them very early in the spring. They will come up when they want to. If you think the weather might be too warm for germination, you can put the seeds in a dish, cover them with water, and keep them in the refrigerator a day or so before planting.

1. Pea vines like to have something to climb on. In your small garden, you could plant them around a tomato cage or a cone made of sticks, depending on how tall your seed package says the vines will grow.

2. Plant pea seeds two inches deep (5 cm) and four inches apart (10 cm).

3. To avoid pulling the vines out of the ground, harvest pea pods using scissors. There are three kinds of peas:

- **Peas to shell** should be harvested when the peas have filled the pods. But the peas should not be so big that the pods won't pop open when squeezed at the tip. If you wait too long, their sugar will have changed to starch, and the peas won't taste as good. Don't eat the shell of this kind of pea.

- **Sugar snap peas** should be harvested when the pods are full-sized and the peas inside are large. Eat whole pods, including the peas, raw.

- **Snow peas** should be harvested when the pods are flat and before the peas inside have started to show as bumps on the pods. Eat whole pods raw or cooked.

Carrot seeds can be planted directly into the garden in early spring.

1. Soak seeds several hours or overnight.

2. Plant seeds 1/4 inch deep (6 mm) and 1 inch (2.5 cm) apart.

3. If your soil tends to get a crust on top, cover seeds with sawdust or sand instead of soil.

4. Keep soil moist but not soggy until all the seeds have sprouted.

5. When seedlings are two inches (5 cm) tall, thin plants to two inches apart by snipping off unwanted ones at ground level. Thinning takes out plants that may crowd each other and leaves space for the remaining ones to grow bigger.

6. As the carrot roots get big, thin the crowded ones by carefully pulling some of them out of the ground. You can eat them!

7. Be sure to keep carrots weeded, especially when the plants are small.

8. Give carrots plenty of water as they grow. But near the end of their growth, too much water may make the carrots split underground.

9. Harvest carrots as you need them but before the ground freezes.

Perennials

Do you remember that perennial plants are plants with roots that live from year to year but the tops die and regrow? Perennials are usually planted in the spring. Your small garden probably won't have room for perennial garden plants, but you may like to grow them in another area sometime. You will need patience to grow asparagus, strawberries, and raspberries. You won't be able to eat them during their first year!

Asparagus is the first vegetable of the spring. Suddenly, when spring has barely

Asparagus spear and a pest called a spotted asparagus beetle. Since adult beetles spend the winter in hollow stems, clean away asparagus branches once they dry.

Raspberries

come, delicious stalks come up out of the ground from the roots. If you plant asparagus, you need to wait three years before harvesting while the plant uses photosynthesis to build its roots. After three years, you can harvest anything thicker than a pencil. Leave thinner stalks to grow and make food for the roots. Pick asparagus stalks by breaking them off near the ground.

Strawberries are a little confusing to grow, but they are so delicious you might want to try some. There are many varieties of strawberries with different traits. You should check with a plant nursery to ask about the best kind for your area. Then, you will need to find out the best way to grow that kind.

There are three groups of strawberries: those that bear fruit once a season, those that bear twice a season, and those that bear a little fruit all season long. Each group has its own instructions for the best way to grow its strawberries. But all three kinds will produce more berries in the future if you pick off all the flower buds the first year. Without flowers, you won't get berries, but more energy will go to strengthening the plant.

Raspberries also come is several varieties that have different instructions for growing. Each plant lives only two years, but a raspberry patch can live for many years.

New raspberry plants sprout from old ones. There are two main kinds of raspberries. One kind grows berries on new plants that grow that year. The other kind grows berries on plants that grew the year before. Check with a plant nursery to see which kind is best where you live. There are different ways to prune raspberries, depending on which you plant.

Roses from a flower shop don't have the beautiful fragrance a garden rose has. Wherever you live in the world, there is probably at least one kind of rose you can grow.

Growing roses is easy. As with strawberries, you will need to find the best kinds of roses for your area. Roses can live many years and are a joy to take care of.

You can buy a rosebush as a bare root or as a potted bush. A bare root will cost less. Either way, dig a hole deeper and wider than the root. Add compost and rose fertilizer to the bottom of the hole first and mix it. Hold the bush in the hole as you add soil back. If you live in a climate that gets freezing temperatures, plant the crown of the bush (the part where its roots come out of the main stem) two inches (5 cm) below ground level. When you cut a rose for a bouquet or prune the bush, cut the stem at an angle just above a stem with five leaflets.

Cut or prune roses just above a stem with five leaflets.

Sniff! Mmmm. I hope you can find a rose to smell this summer!

Prayer

Father, we see Your care for us in the food and flowers You have provided for our gardens. You made different plants that will give us food to harvest at different times. You made roses, the beautiful flower everyone enjoys, grow in all the climates where people live. Thank You for garlic's flavor and healthy phytochemicals. Thank You for purply-red radishes and their surprising flavor. Thanks for fresh mild lettuce, peas from the vine, asparagus, and sweet berries! Amen.

CHAPTER 35

Sow and Grow Summer Produce

> When you eat the labor of your hands, You shall be happy, and it shall be well with you. (Psalm 128:2)

I hope you work hard in your garden. God says that eating the food from your garden will make you happy!

Summer is when the garden explodes in fruitfulness. Sometimes it's even hard to see any dirt between your garden plants! By midsummer you won't need to weed as much because your crops will be large enough to shade the ground and discourage new weeds. Let's learn about a few kinds of summer produce God has provided.

Tomatoes

Tomatoes aren't like some plants that need certain temperatures to bloom and make fruit. Their plants just need to be old enough. But tomatoes can't handle cold temperatures like the early spring crops can. To get the most produce from tomato plants, you can plant seeds indoors in early spring and **transplant** them to the garden later. This will give them a head start. Or you can buy tomato plants from a nursery in late spring when the air and soil is warm enough for tomato plants to be planted outside. Here's how you can raise your own tomato plants indoors from seeds:

1. Collect small containers like yogurt cups or the plastic packs that bedding plants come in. Collect the same number of containers as the number of tomato plants you would

4. Poke 1/2-inch-deep (1 cm) holes in the soil of each container.

5. Drop two tomato seeds in each hole.

6. Bury the seeds with soil.

7. Cover the pots with plastic wrap and put them in a warm place (like on top of the refrigerator). Light is not important until the seeds come up.

8. When the seeds come up, remove the plastic wrap and place the pots in a sunny window. If you have a grow light, you can shine it on the plants 12-14 hours a day, even while the

like, plus one or two extra.

2. Poke drainage holes in the bottoms of your containers.

3. Fill the containers with moist potting soil.

Planting tomato seeds

Tomato plants indoors with sunlight and full spectrum fluorescent light.

sun shines. This will help the plants grow better. Keep the light two inches (5 cm) above the tallest leaves by raising the light or lowering the plants as they grow.

9. When the seedlings are two inches (5 cm) tall, cut off the weaker of the two at ground level with scissors.

10. Water every day or two. Don't overwater.

11. Rub your hand across the tops of the plants for a few minutes several times a day. This makes the plants think they are being blown by the wind, and their stems will become stronger.

12. When seedlings are about six inches tall (15 cm), transplant each one to a larger container like a 32-ounce yogurt cup with drainage holes. Snip off the plant's lowest set of leaves. Put the plant deep in the cup so the soil comes up to the leaves that are now lowest on the stem. The plant will sprout more roots along its stem!

Move your tomato plants outdoors:

1. Leaving your plants in their pots, gradually get them used to being outdoors. When all danger of frost is past, you can begin to move your tomato plants outdoors. Over

Cut off the lowest two leaves whenever you transplant tomatoes.

a week's time, set them in the sun longer amounts of time each day. Start with two hours and end with all day. Leave them out at night too.

Early summer garden with lettuce bed and tomato plants in protectors

2. To transplant your tomato plants into the garden, remove the lowest set of leaves as before. Plant the plants as deep as the leaves that are now the lowest on the stem.

3. Carefully poke a small rough stick into the ground next to the stem. Cutworms normally like to wrap themselves around stems while chewing on them, but the stick discourages them.

4. Put a tomato cage around each plant for support.

5. To protect new plants from too much heat, cold, or wind, you can buy little shelters to put around each cage. These plant protectors are made of connected plastic tubes that you fill with water. They even protect against frost so you can plant your tomatoes early. Plants grow more quickly with these warming shelters. Be sure to remove the protectors before your plants start spilling over the top.

6. Enjoy your harvest all season long as the tomatoes ripen.

7. Many tomato plants keep producing fruit as long as they are alive. But they will freeze easily in the fall as temperatures dip. You can cover your plants with blankets if there is an early frost predicted. Uncover them when the weather improves. Don't cover tomato plants with plastic since plastic doesn't shield from cold. Before you think your plants will freeze even under their blankets, pick the green tomatoes. They will gradually ripen, unrefrigerated, indoors.

Beans and Corn

Beans are easy to grow. Beans that you eat fresh (like green beans) don't take long to grow. Beans to dry (like pinto beans) take longer because they need to dry on the plant.

1. Plant bean seeds one inch deep (2.5 cm). Some bean plants are like little bushes, and others are vines. Check your seed packet for instructions on the amount of space to leave between seeds for your variety.

2. To keep bean seeds from rotting in the ground before they sprout, water once with a good soaking right after planting. You won't need to water the seeds again until they sprout unless the weather is very hot and the soil becomes dry at the depth of the seeds.

3. Pick beans when they are a good eating size but before the seeds start to bulge in the pods. If you keep picking the young beans, the plant will be encouraged to make flowers and grow more beans.

Corn can be planted directly into the soil when all danger of frost is past or, as folks used to say, when the cottonwood leaves are as big as a mouse's ear.

Do you remember that the tassels of corn are the pollen-making flowers, and the corn silk is the part that takes the pollen? And do you remember that corn pollen is spread by the wind? These are good reasons to plant corn in a square rather than in a row. Keeping the plants together helps pollen land where it should.

1. Plant corn seeds 2 inches deep (5 cm) and 18 inches apart (45 cm) in a square bed.

2. Ears are ready to pick when the silk is dark brown but still damp. The tops of the ears should look rounded in their husks, not pointed. Carefully open a husk and press a kernal wiith your fingernail. If it easily squirts, the corn is ready to pick! If the kernal is tough and hard to puncture, you have waited too long.

The sooner you eat the corn after it's picked, the better it will taste. Some

Zucchini

people even get a pot of water boiling, then pick the corn, shuck it as they run back to the kitchen, and drop it in the pot for five minutes. Be sure to have butter and salt ready too!

Cucumbers and **zucchini** are good summer crops you may want to grow. We will talk about how to plant them in the next chapter when we learn when we learn about planting later crops. Both cucumbers and zucchini are best if picked before their skins get tough and their seeds get large and hard. Some kinds of cucumbers are better for pickles, and other kinds are better for salads.

Zucchini fruits are sneaky. They are so good at hiding that it's easy to miss one until you trip over it! Try to pick zucchini when it's six to ten inches long (15-25 cm). Zucchini plants are so large that one plant could fill a fourth of your plot.

Flowers

You will show me the path of life; In Your presence is fullness of joy; At Your right hand are pleasures forevermore. (Psalm 16:11)

Flowers are some of the pleasures God gives us!

It's fun to add some color to brighten up the greenery in a vegetable garden.

Marigolds: picking annual flowers and cutting off dead ones encourages more blooms.

Flowers are the most beautiful way to do that. Flowers can do helpful jobs too! Here are a few easy-to-grow flowers to plant in your garden.

Marigolds are brightly-colored annual plants that are easy to grow. Marigold flowers are not only cheerful looking, but they also have many jobs they can do for you:

- Marigold flowers are edible and contain healthy phytochemicals. Sprinkle a few petals on a salad!

- Slugs like to eat marigold leaves. Slugs may leave your lettuce alone if you plant marigolds near your lettuce.

- Marigold flowers attract pollinators that are beautiful (like butterflies) and useful (like the bumblebees your tomato blossoms need).

- The strong smell of marigolds may help discourage pests from coming into your garden.

Far away from your garden, marigolds are raised for other uses. They are harvested for medicine, health, fragrances, food dyes, and pest control.

Zinnias are showy flowers with lots of colorful petals. They attract pollinators and are easy to grow! It's easy to save zinnia seeds, so you might want to plant heirloom varieties.

1. Plant seeds 1/4 inch deep (0.5 cm) directly into the garden once the soil is warm. Or you can plant seeds indoors up to six weeks early for a head start. Space them according to the seed packet instructions for your variety.

2. Seedlings should come up in about a week!

3. During flowering, you can trim off dead blossoms to encourage more blooms. But if you want to save seeds for next year, leave a few flowers to dry.

Prayer

Father, we thank You for plentiful zucchini we can share with others. Thank You for bright, helpful marigolds and friendly zinnias. It's so fun to bite into a warm, bursting tomato out in the garden. We praise You for steaming corn on the cob as it melts its butter. And thank You for crisp cucumbers and bountiful beans. Amen.

CHAPTER 36

Sow and Grow Fall Produce

Fall is coming! As your summer harvest slows down and plants start to turn brown, bright orange pumpkins won't be afraid to show their happy colors. Fall is their time to ripen. Sunflowers, onions, and winter squash will also be ripening. These foods have taken a long time to get ready for harvest even though their seeds were planted at the same time as the fast-growing summer vegetables.

You crown the year with Your goodness, And Your paths drip with abundance. (Psalm 65:11)

God is so good to give us generous amounts of food all year long. We especially see how He provides when we harvest in the fall!

Pumpkins and Winter Squash

Pumpkins and **winter squash** are healthy foods we can grow and store to use all winter. Some pumpkins that are grown for size or decoration are stringy. For smooth eating, be sure to get seeds for "pie pumpkins."

1. Pumpkins and winter squash need a lot of space to grow. If you plant them on the edge of your small garden, you can adjust their vines to spread on the ground outside the garden. Don't plan to have the vines spread over concrete because they like to make extra roots as they grow.

2. Before planting, make a hill of soil two feet wide (60 cm) and eight inches (20 cm) high.

3. Plant four seeds one inch deep (2.5

cm) and a few inches apart in the top of the hill. Cucumbers and zucchini can also be planted one inch deep in hills.

4. When seedlings have grown a few leaves, pull out all but two of the healthiest plants.

5. Cut ripe pumpkins and squashes off the vines when their skins have become tough just before the frosts of fall. Leave a few inches of stem on each.

6. Cure pumpkins and squashes for storage by placing them in a dry, warm room for two weeks to harden their skins. To avoid rot, set them on paper or straw, and place them far enough apart that they aren't touching each other.

7. Store in a cool, dark place.

Sunflowers

Sunflowers are fun to grow! Some can grow taller than Goliath and have flowers the size of a platter. It's wonderful to eat the seeds from the flower's center after they dry. It's also fun to leave the dry flowers on the plant and watch birds come to eat the seeds. If you want to grow a large sunflower, you will need a sunny square about two feet by two feet (0.5 m x 0.5 m). Check the seed package to see if you should plant seeds early indoors in your area.

1. Plant seeds 1/2 inch deep (1 cm).

2. In windy areas, you may need to tie the tall flower stems to stakes.

3. Sunflower seeds are usually ready to harvest when the yellow ray flowers dry up and the seeds in the center disk turn brown. Pry out a seed, crack it open, and taste the inside to see if it's ripe.

4. To harvest ripe seeds, cut the flower head off and hang it upside down in a dry place indoors.

5. After two or three weeks, carefully remove the seeds with your fingers or a fork.

Prepare for Pests and Plant Diseases

Growing a garden is fun! It's a joy to work hard the way God designed us to do. And it's a blessing to eat the things we grow. But sometimes we have garden problems. It's safer and healthier to use natural ways to solve these problems instead of using chemical sprays. Let's see what some of these garden problems might be and learn some safe ways to fight them.

 Problem: Creatures that eat our produce before we do. Bugs, birds, and mammals can eat or damage our crops. Having just a few bugs won't hurt your garden. One bug isn't big enough to eat a lot. But it's a good idea to keep a container of water with a few drops of dish soap per quart (liter) mixed in. You can pick or knock off problem bugs into the water, and they will quickly drown.

A bigger problem is when a bird, squirrel, raccoon, or deer comes to eat or ruin a lot of food. For these large animal problems, you may need

Japanese beetles can ruin rose blossoms.

Drowned beetle pests

fences around the garden or netting over certain plants.

 Voles (above) are small rodents with thick fur coats and short bare tails. They can eat their weight in greenery (like these green beans) every day! You can use mouse traps baited with nuts to trap them. Allow friendly snakes to live in your garden.

Grasshoppers can damage your garden when there are a lot of them hopping, flying, and chewing. Otherwise, they are fun to catch! This young grasshopper doesn't have adult wings yet—it has wing buds.

The **apple coddling moth larva** damages apples.

Cutworms are moth larvae that chew down plants at ground level in the night.

332

Tomato hornworm **frass**
(droppings)

A ladybug on parsley looks
for its favorite meal of aphids.

Tomato hornworms eat large amounts of tomato leaves, stems, and fruits. They are often hard to find, but their specially-shaped droppings give them away.

God has given us lots of help from insects that eat other bugs. We can encourage these good insects by leaving dead plants in the garden to give them places to hibernate. Get to know these helpful bugs in your garden:

Let's cheer on these garden helpers!

A daddy-long legs looks for bugs to eat on a turnip plant.

This spider stopped a Japanese beetle from its meal of grape leaf.

Assassin bugs pounce on their dinner.

Katydids are usually camouflaged with their leaf-like wings. They eat both plants and bugs.

God protects hover flies from birds by giving them a wasp costume. Hover fly larvae eat lots of aphids!

A praying mantis searches for a tasty bug to grab with its strong front legs. Is she looking at you?

If you see eggs like these above, let them be. They belong to the insect-eating green lacewing.

Problem: Weeds that use up space, soil nutrients, and water meant for our gardens. Any plants that **compete** for things our plants need are called weeds. Even garden plants that grow where we don't want them are weeds. The hardest weeds to get rid of are perennial weeds with long branching roots that break off underground when you try to pull them up. A new plant can come back from each piece of broken root.

Pull weeds as early in the season as possible, but especially before they spread their seeds. If a weed is near a garden plant, you may want to cut or hoe it down instead of pulling it. Pulling may damage the roots of your garden plant.

 Problem: Plant diseases that make your plants sick. Plant diseases are sicknesses caused by microscopic things like bacteria, viruses, and fungi. If you notice something wrong with a plant or part of a plant, you should remove it and throw it away. Don't put the sick part in the compost bin.

You can prevent some diseases by leaving space between plants. This allows air to carry away microscopic problems instead of letting them move from one plant to another.

Early in the season, wasps eat bugs like this caterpillar. Later, they become pests by eating sweet fruit.

Thistles are perennial weeds that can spread by roots and seeds.

Dandelions can grow from the grasslands to the mountains above timberline. We often think of dandelions as weeds, but their plants have value for food and health.

Keeping your garden healthy will prevent some diseases. Make sure your plants have enough but not too much water, and keep your soil nutritious with compost or fertilizer. You can also take a sick piece of plant to a plant nursery for advice. Be sure to put it in a plastic bag so you don't spread disease to their plants.

Prayer

Lord, thank You for crowning the year with Your goodness. We thank You for the garden food we can store through the winter. We will thank You all year long for Your goodness! Amen.

We have loved learning about plants with you! As our beautiful, fruitful gardens grow, let's remember Psalm 90:17!

"And let the beauty of the LORD our God be upon us, And establish the work of our hands for us; Yes, establish the work of our hands."

Image Credits

Most images are taken from iStock.com, with the following exceptions:

Chapter 1

1. Herby and Flora characters — Tamela Sechrist
2. "Cutting hay" — John Schneider
3. "Round hay bales" — John Schneider

Chapter 5

4. "Backyard bouquet" chapter picture — Tamela Sechrist
5. "Mariposa lily" — John Schneider
6. "Bird of paradise flower" — John Schneider
7. "Passionflower" — John Schneider
8. "Apple blossoms" — Tamela Sechrist
9. "Sunflower head" — John Schneider
10. "Tomato blossoms" — Tamela Sechrist
11. "Blackberry blossom and pollinator" — Tamela Sechrist

Chapter 6

12. "Sticky geranium" chapter picture — Dr. Neal Bringe
13. "Melissa blue butterfly on yellow salsify" — Dr. Neal Bringe
14. "Pollen-making squash flower" — Tamela Sechrist
15. "Pollen-taking squash flower" — Tamela Sechrist
16. "Tulip" — Tamela Sechrist

Chapter 7

17. "Honeybee" — Dr. Neal Bringe
18. "Pollen wasp" — Dr. Neal Bringe
19. "Blooming yucca" — John Schneider
20. "Yucca moth" — Dr. Neal Bringe

Chapter 8

21. "Desert soil" — John Schneider
22. "Desert under cliffs" — John Schneider
23. "Saguaro" — John Schneider
24. "Cactus spines" — John Schneider
25. "Shrunken cactus" — John Schneider
26. "Swollen cactus" — John Schneider
27. "Sand verbena" — John Schneider
28. "Pollinator on prickly pear" — John Schneider
29. "Collared lizard" — John Schneider

Chapter 9

30. "Succulents" — Dr. Neal Bringe
31. "Parts of a Leaf" diagram — Tamela Sechrist
32. "Skunk cabbage" — John Schneider
33. "Red aspen leaves" — John Schneider
34. "Ash tree leaflets" — Tamela Sechrist

Chapter 12

35. "Child holding red leaf" — Tamela Sechrist
36. "Aspen forest" — John Schneider
37. "Winter oak leaves" — John Schneider
38. "Red aspen branch" — John Schneider
39. "Dogwood blossoms" — John Schneider
40. "Redbud blossoms" — John Schneider
41. "Trillium" — John Schneider
42. "Fern" — John Schneider
43. "Phlox" — John Schneider

Chapter 13

44. "Red shrub" — Tamela Sechrist
45. "Mountain wild flowers" — John Schneider
46. "Grape tendril" — Tamela Sechrist

Chapter 32

108. "Sea falls" — John Schneider
109. "Light rays on the sea" — John Schneider
110. 'Smallest photosynthetic thing" — Wikimedia Commons
111. "Phytoplankton varieties" — Wikimedia Commons
112. "Father seahorse" — Wikimedia Commons
113. "Sea urchins" — Wikimedia Commons
114. "Sea otter with urchin" — Wikimedia Commons
115. "Crab eating a sea urchin" — Wikimedia Commons
116. "Krill" — Wikimedia Commons
117. "Gull and sea lions" — John Schneider

Unit 9 Introduction

118. Dr. Neal Bringe

Chapter 33

119. "Making a garden plan" — Tamela Sechrist
120. "Worms in compost" — Tamela Sechrist
121. "Unfinished compost" — Tamela Sechrist
122. "Finished compost" — Tamela Sechrist
123. "Garden plans" — Tamela Sechrist

Chapter 34

124. "Garlic scape" — Tamela Sechrist
125. "Lettuce seeds" — Tamela Sechrist
126. "Opening pea pods" — Tamela Sechrist
127. "Asparagus spear and asparagus beetle" — Tamela Sechrist

128. "Raspberries" — Tamela Sechrist
129. "Front yard roses" — Tamela Sechrist
130. "Pruning roses in summer" — Tamela Sechrist

Chapter 35

131. "Planting tomato seeds" — Tamela Sechrist
132. "Indoor tomatoes with sunlight and full spectrum bulb" — Tamela Sechrist
133. "Cutting leaves when transplanting tomatoes" — Tamela Sechrist
134. "Early summer garden" — Tamela Sechrist
135. "Zucchini" — Tamela Sechrist

Chapter 36

136. "Japanese beetle in rose" — Tamela Sechrist
137. "Drowned beetles" — Tamela Sechrist
138. "Apple coddling moth larva" — Tamela Sechrist
139. "Young grasshopper" — Tamela Sechrist
140. "Tomato hornworm frass" — Wikimedia Commons
141. "Ladybug on parsley" — Tamela Sechrist
142. "Daddy long legs on turnip" — Tamela Sechrist
143. "Spider eating a Japanese beetle" — Tamela Sechrist
144. "Katydid" — Tamela Sechrist
145. "Praying mantis" — Tamela Sechrist
146. "Assassin bug" — Wikimedia Commons
147. "Dandelion" — Tamela Sechrist

Notes

1.   "The Origin of Oil," www.answersingenesis.org/geology/the-origin-of-oil/.
2.   "10 Surprising Facts About Grass", https://www.pennington.com/all-products/grass-seed/resources/10-surprising-facts-about-grass
3.   "How Far Can the Human Eye See?," https://www.healthline.com/health/how-far-can-the-human-eye-see
4.   "The Fantastic Fur of Sea Otters," https://www.kqed.org/science/25908/the-fantastic-fur-of-sea-otters